GW01246732

The Second Life of Networks

Didier Lombard
with Georges Nahon
and Gabriel Sidhom

The Second Life of Networks

Copyright © 2008 by Odile Jacob Publishing Corporation, New York.

www.odilejacobpublishing.com

All rights reserved. Printed in France.

No part of this book may be reproduced in any form or by any electronic or mechanical means, including storage and retrieval systems, except in the case of brief quotations embodied in critical articles and reviews.

FIRST EDITION

Printed on acid-free paper.

Library of Congress Control Number: 2008931838.

ISBN : 978-0-9768908-1-2

Acknowledgments

Above anything else, this book is the result of daily reflection while working along the 190,000 employees of the France Télécom-Orange Group who, by their work, creativity, and ability to adapt make it possible for the Group to transform itself and change the world around it. My sincere thanks go to them first.

Of equal importance was the knowledge and inspiration I obtained from my numerous interactions with the customers and partners of France Télécom-Orange. The one thing that unites everyone, whether in large cities or small villages, either in the business or very far from it, is a clear interest and vivid curiosity regarding the new world of second-life networks.

Next I would like to sincerely thank Elie Girard and Jean-Paul Maury for their invaluable contribution throughout the writing of this book as they encouraged me to go beyond what I had imagined.

Finally, I would like to thank Vivek Badrinath, Olivier Barberot, Nicolas Guérin, Caroline Mille, Pascal Périn, Mark Plakias, and Raoul Roverato from France Télécom-Orange and, outside the Group, Jean-Louis Gassée and Dominique Nora, for taking the time to read the book and provide insightful comments.

All royalties generated by this book's copyrights will be donated to the Orange Foundation.[*]

Created in 1987, the Orange Foundation is a nonprofit philanthropic organization whose aim is to provide everyone with the means to communicate in the broadest sense. In addition to the spoken word, the foundation is involved in major battles against autism, visual and hearing impairments, and illiteracy, and also supports vocal music.

* http://www.orange.com/en_EN/corporate_philanthropy/foundation/

Foreword

The late Roy Amara's First Law of Technology[*] states: "With every change in technology that affects consumer behavior, we always overestimate the impact in the short term, but then underestimate the full impact over the long term."

In the last fifteen years, in networks and computing, many predictions on a technology's uptake got either the intensity or the timing, and sometimes both wrong. This led many to become skeptical and lose faith in the promises of technological innovations, especially after the burst of the telecom and Internet bubble in 2000.

Early on, visionaries predicted that people throughout the world would one day flock to a service where individuals could communicate across television-enabled telephones. It is a legend often repeated today that although precisely such a service, called the Picturephone, was invented in 1964 by AT&T's Bell Labs, it did not go anywhere. The lack of interest in such a service by individuals was confirmed repeatedly over the subsequent decades and across numerous countries. It seemed as if the service could not even be given away. But eventually a conjunction of circumstances helped make the service popular, though in a different way than originally imagined. First, the Internet became a worldwide web. Second, it

---

* http://en.wikipedia.org/wiki/Observations_named_after_people

became possible to communicate with anyone in the world for virtually the same low price independent of distance and duration. Innovators had suddenly found a way to leverage the computing power of the PC, and the ability to connect to the Internet (albeit at slow speeds compared to today) to introduce a new form of video telephony. Except that this "video telephony" consisted of video on the Internet and not on the telephone.

According to Anjuili Elais, "In 1994, San Francisco State students Jeff Schwartz and Dan Wong constructed a device called a webcam, which captured snapshots every minute and broadcast them in real time on the Internet. Today that webcam, nicknamed the SF State Fogcam, holds the distinction of being the world's oldest webcam in operation."*

So the webcam emerged and began to be manufactured in great numbers even though people did not use them as frequently as imagined. Moreover, the generation using the webcam was several steps removed from the one that initially invented the Picturephone. Does this mean that all innovations appearing before their time will always succeed at some later time? Probably not. There is a generation gap, and services conceived by members of one generation for those of subsequent ones rarely pan out. In many cases, these one-generation-removed innovations are adopted by later generations in illogical and nonlinear fashion. Indeed, the members of the new generation differentiate themselves from their "parents" in the very way they use these services. They create a kind of counter-culture based on consumption patterns. For technically oriented individuals, it becomes a conscious personal challenge to push beyond the established rules that define how a service or device is to be used. There are many motivations for this behavior, ranging from the hacker who does not want to obey the "rules" associated with the service or device, to the tinkerer who simply believes there is a better way to conceive them. All this is governed by some overarching

---

* http://xpress.sfsu.edu/archives/life/001905.html

concerns: to try and do more with less; to stretch the frontiers of the existing technology; and to repurpose technology to do something different than what it was originally intended for. These individuals are not predisposed to accepting the given or to simply improve it incrementally. It is precisely this attitude, both audacious and irreverent of the legacy of the past that leads to unexpected innovations. Today, networks accelerate the adoption of these attitudes because they allow quick, persistent, and constant collaboration among millions of people throughout the world. In addition, the amount of information available on the Internet is enormous, ever growing, and updated by the minute through significant contributions by its users. As a result, what can be changed will be changed because the means to do so are at everyone's fingertips. But it is not enough that a technology improves people's comfort, that it saves money or time, or that it offers entertainment. This is table stakes for any technology. It also needs a bit of magic, according to the late British science fiction writer Arthur C. Clarke.* His third law of technology states:** "Any sufficiently advanced technology is indistinguishable from magic."

People tend to explain what they see by extrapolating from earlier versions of similar concepts. Innovation cycles simply churn things, whether or not the new developments represent an improvement in speed or quality, and whether or not they have any immediate relevance. For example, nobody today would tolerate navigating through Internet sites as they were five to ten years ago. The same is true for mobile phones whose styles and features keep changing, making previous models obsolete.

In the techno-economic literature, there are many references to statements called "laws." These laws purport to summarize empirical observations of past trends by functionally reducing them to an expression that allows for extrapolation—the projection into the

---

* Clarke died, at age 90, in March 2008.

** http://en.wikipedia.org/wiki/Clarke's_three_laws

future of past tendencies or identification of correlated factors. We have already referred to two such laws.

In a world of growing abundance and variety of information, there is a need for beacons to mark and summarize what is experienced. On the Internet, there are fewer established, "de jure" authorities and more "de facto" bodies of opinions and recommendations. Networks, through their speed, ubiquity, and the persistence of connected services, have allowed new modes of use. For example, there is the new concept of "living on the Internet." This has given rise to the expression "life-feed" to describe services provided by companies such as Twitter, Facebook, and Plaxo. The formulation of "laws" of technological development helps encapsulate such change in real time, enabling future projections.

One law that is not frequently mentioned is Shannon's law. It was popular in speeches, conferences, and articles in the 1970s and 1980s. Shannon's law, formulated by Claude Shannon,* states that there is an upper limit to the data throughput that can be obtained from a given transmission channel. It was developed at a time when network bandwidth was scarce and could not carry the level and type of traffic we see today. The laws quoted today reflect the reality of abundant bandwidth. They tend to give an exponential vision of the evolution of the technology business. Examples include Moore's law and Metcalfe's law (see Chapter 3), which are structurally optimistic. Their underlying mathematics allows no foreseeable downside in the future of information and communications technology.

The Internet and its constituent networks have developed in a very short period of time when compared to other technologies and inventions. Today, we have already reached the limits of this relatively new space. Many companies have taken to growing by merger

---

* Claude Elwood Shannon, a mathematician born in the U.S. in 1916, developed the new field of *information theory*. Shannon spent most of his career at Bell Labs where he worked from 1941 to 1972 as a mathematician dedicated to research. He died in 2001. (http://whatis.techtarget.com/definition/0,,sid9_gci214303,00.html)

and acquisition instead of organically. Most of what needs to be done to expand the frontiers of the Internet and its networks is either possible or imaginable. What is missing is well understood by the larger community of technologists, business people, regulators, and lawmakers. Paradoxically, most initiatives that emerge from this complex ecosystem end up being linear and incremental extrapolations of the past, an approach typical of networks' first life. Innovation moves so fast that the political and business support system needs to be extremely agile.

Today there is a new wave of innovative ideas; scientists, engineers, and "tech" entrepreneurs have seen the promise of a future beyond the limits of the current networks and information technology. Actually, some of the more forward thinkers have already gone past these frontiers and reached new areas; this is the case with "green" and "clean" technologies. The technological status quo is being challenged, and this will result in more sustained and genuine innovation. Past ways of thinking cannot give us a model of the unknown world before us. Only a more unassuming but nevertheless audacious and nonlinear approach will do.

Only along this expanded frontier can ideas and businesses appear spontaneously once again and go on to challenge the existing players. Companies must constantly challenge the comfort zone created by their past and current successful operations. Securing the younger generation's approval to and their participation in a company's vision has become a key obligation to stockholders. These younger people represent a major source of influence and agility, both necessary factors to build a technological future.

Abandoning the current technological incrementalism will require significant investments to unleash fresh disruptive innovations. High-capacity network, until now available only in the backbone will have to be deployed closer to customers. Intelligent infrastructures that can adapt to the level and type of traffic will also be needed. We can gain a glimpse of what is ahead by looking at the sophisticated datacenters conceived and built by Internet giants such as Google, Yahoo!, Microsoft, Amazon, and eBay. These new networks and technologies are tools that can be used for many purposes. Vigilance will be necessary to ensure that illegitimate uses of communication media, such as those that have emerged in the past, are kept in check.

*

It seems today that almost any technology concept is modified by escalating numerical suffixes (such as 2.0 or 3.0) to represent its "next big thing" potential. One could argue that referring to the second life of networks is just a more subtle way of doing the same. In what follows, I will certainly try to avoid this headlong rush into "suffix faddism."

Nevertheless, the current profound technological changes are not just impacting a small circle of "techno-elites." Rather they concern every one of us. I have spent numerous years as a direct or indirect part of what has become the France Télécom-Orange Group, of which I am proud to be now Chairman and CEO. All of the knowledge and experience that I acquired during those years have convinced me that we are seeing today the dawn of a fundamental shift in the life of networks. In this revolution, "physical" networks will be totally fused with virtual, social, and human networks. But they will of course remain the irreplaceable pedestal of our communications environment.

Read this book. Read it to its end. You will end up asking yourself if sometimes trees do not grow all the way up to the sky.

Didier Lombard
May 2008

Introduction

One hundred and thirty years after the invention of the telephone, which marked the dawn of what we will call the "first life of networks," the world is everywhere connected to a multitude of interwoven wire-line, wireless, human, and social networks.*

It is hard to imagine that the inventors and network operators present during the early years of telecommunications could have foreseen such profound and rapid changes in their industry.

1975-1995: Three "Big Bangs"

By the early 1970s, communication networks throughout the developed world had undergone a metamorphosis. This was part of the rich fallout from a series of three technological "Big Bangs": the inventions of digital technology, the Internet, and mobile wireless communications.

---

* The URL references provided point to Internet page addresses containing content that were still available as of June 2008.

1995–2004: First Steps Toward the Second Life of Networks

The twenty years marking the transition between networks' first and second lives might have led to an impasse, had they not been followed by a decade of widespread propagation of these three technological breakthroughs.

This phase saw the invention and deployment of Internet browsers (chief among them Netscape's Navigator),[1] the only tools capable of providing simple, ergonomic access, appropriate for mass-market consumers, to the already considerable amount of content available on web sites. This was followed shortly by the development of search engines (such as those of Yahoo! and Google), which allowed users to focus the range of sites they were accessing.

At about the same time, fueled by falling costs and usability improvements, mobile phones began to be widely distributed, first among business users, then more generally among consumers. This period also saw the development of original and unexpected types of usage of these technologies.

The new generation of users, popularly called "GenY" or the "Net Gen," has transformed networks into a melting pot of services and content. It is therefore only natural that a new generation of entrepreneurs has emerged to defy the status quo and overcome the hierarchical constraints of existing technical architectures.

The modification of the telecommunications value chain represents a profound change in the sector's industrial organization. Established operators have had to share access to their markets with new, alternative operators who emerged following the deregulation of the industry. New segments of the value chain were created, occupied by the then embryonic online service and content providers, some of them newly created, some coming from other mainstream sectors such as the media sector.

This period begins with the IPO of Netscape and ends with the one of Google in the summer of 1995 and 2004, respectively. This interval represents the "gestation" period for the second life of networks.

At the beginning of the third millennium, information and communication technologies are no longer restricted to a small group of the fortunate technophile elite; rather, they are broadly available and accessible to the majority of consumers.

The necessary ingredients are finally in place to catalyze both the proliferation of novel uses of the technological fruits of the previous decade and the emergence of social networks that began to blend themselves into available communications infrastructures. This marked the beginning of the second life of networks, whose services have become a permanent feature of contemporary life.

The User Becomes an Active Network Node

The gradual simplification and convergence of technologies have given users a large degree of control and have blurred the distinction between physical and social networks. For example, a user today can generate and distribute content and services as well as recommendations, advice, or guidance through the new capabilities afforded by these intertwined physical and social networks. Users journey back and forth between virtual and real spaces seamlessly, to the point where the two states are sometimes undistinguishable. It is common today for users to go from a social conversation on a mobile phone to an online forum, then to sharing the most recent vacation pictures with their distant family, and finally to navigating a personally configured avatar to a music concert in a parallel virtual world—often all in real time. Users have evolved from being a "cold," mostly passive network endpoint to a constantly active node at the center of a mesh of interconnected networks, pulsating with sent, received, and forwarded information.

The Emergence of the "Free" Business Paradigm

Along with the second life of networks came a new business model, one based on the elegant and powerful but complex principle of a price perceived to be "free" by the end-user. In this model, online content and service providers tap a new and growing revenue stream coming primarily from online advertising.

What today appears obvious was not at all evident at the time. It seems to have resulted instead from a combination of circumstances. First, the Internet was able to penetrate mass markets beyond those limited to academic and computer-literate users. This was achieved at the beginning with dial-up modems, then with "always-on" ADSL services operated by telecommunications operators. In the case of the U.S., cable modem service operated by cable TV companies emerged as an alternative to ADSL.

Second, in this model, ADSL and cable access was made possible by using the underlying physical networks—copper for telephone lines and coaxial cables for cable TV—of the first life of networks. These infrastructures were already globally amortized in most developed countries. Further, the overinvestment in backbone network[2] capacity occasioned by the Internet/telecom bubble of 1995–2001 accelerated the investments in networks. At the macroeconomic level, the growth of the ad-supported business model on the Internet is largely dominated by globally operating US-based service companies which has led to a financial transfer from Europe and Asia to the United States.

The "Amalgamation" of the Value Chain

At about the same time, telecom operators underwent a transformation evidenced by an amalgamation in the value chain they occupied. Increasingly, all types of players, old and new, operators or equipment manufacturers, service as well as content providers, are venturing into the emerging territories at the edges of their core businesses.

One example of this trend is the recent acquisition by equipment and device manufacturers of service companies; another is of services companies themselves participating in wireless spectrum auctions. There are several reasons why this is happening. First, players are seeking to establish and grow a direct relationship with their customers. Second, the advertising-supported business model is forcing the previous generation of players to shift part of their businesses toward this new source of revenues.

There is also a growing tendency for some service companies to control their own communications and datacenter infrastructures. Some are investing heavily, attaining levels equivalent to those of established network infrastructure companies.

One other aspect of the trend is the proliferation and growing complexity of available services made possible by advances in information technology. Getting users to adopt and actually use these new services requires the ergonomic simplification that convergence allows. This encourages companies to partner in order to provide complete and compelling experiences for users.

Finally, the innovation process has evolved gradually from a "cathedral" model, in which a company develops virtually all of its products and services internally or through suppliers, to one characterized as a "bazaar,"[3] where a company primarily integrates third-party companies' service innovations. The bazaar model often results in alliances between companies, strategic partnerships, or the acquisition of leading innovative companies.

Fully Embracing the Second Life of Networks

The transformation we have seen among key players in the communications, information, and media ecosystem is probably a sign of acceleration in the development of the second life of networks.

The ability to provide ubiquitous access to social networks and associated services is, however, facing some obstacles. Key among them is the physical limitation of the copper access infrastructure inherited from networks' first life. Exciting and potentially transformative services, such as high definition and 3D television are appearing on the horizon, but they will be unable to develop their full potential with the limited capabilities inherent in first-life networks. The significant increase in bandwidth required to power such services will only be possible with fiber optic networks (about 100 megabits per second, or ten to twenty times higher than currently available speeds). There will also be improvements in the speed of mobile networks, as they evolve to so-called 4G (Fourth Generation) or LTE (Long Term Evolution). These two trends in fixed and mobile networks will open the way to the core of the second life of networks.

In the case of fiber, even more significant is the balanced receiving and transmitting bandwidth speeds, which will allow for symmetrical exchanges of information. This is an important prerequisite for the user to become both a producer and a consumer of content. Social networks will evolve to their next stage, which will be characterized by personalization and mobility. This continuous management of preferences and context, geographic and temporal, will allow users to discover "hyper relevant" services delivering increased personal and professional convenience and efficiency.

As Paul Otellini, CEO of electronic chipmaker Intel, remarked during an interview with *Newsweek*'s Steven Levy:[4] "We now have a 'go-to' Internet. You go to search for something, you go to a Web site, you go to buy something. I think a better model would have the

Internet come to you. The Internet would be interpreting things for you, translating signs for you, giving you directions when you need them, as opposed to when you ask for them... If the Web becomes this more immersive environment, it will need to have a lot more intelligence. It will require a tremendous amount of processing power to make it real."

We are currently experiencing the growth phase of the second life of networks. This represents a new and exceptional opportunity for human and social progress, one at least as important as that occasioned by their first life. It can also contribute to sustaining and reinforcing the developed world's economies. It is unlikely, however, to do so by itself without other factors, chiefly associated with public policy, coming into play.

From a microeconomic standpoint, the deployment of fiber networks is based on the assumption that there are clear and stable "rules of the game" providing a basis for predicting return on investment. This will allow investors to commit funds for financing such significant projects. This time around, the underlying networks are not amortized, given they are at the beginning of their deployment, and will therefore require extensive capital investment.

From a macroeconomic standpoint, the fact that most online advertising revenues from Europe, and Asia accrue to a small number of international U.S. companies should spur worldwide innovations by service and content companies in Europe and Asia.

Finally, one of the remaining social challenges, requiring both technological and behavioral evolution, is the protection of individuals' personal data.

Clearly, in order to assess the stakes, opportunities, and challenges that second-life networks represent, it is important to understand their genesis. This is the objective of this book.

CHAPTER 1

Once Upon a Time, in Networks' Earlier Life...

The history of telecommunications is about coming back full circle, re-discovering as it were, natural principles that were surrendered in the myriad compromises made to accommodate new technologies. We will shortly delve into those aspects of the history of telecommunications which are important for understanding how second life networks have developed. Before we do so, it will be useful to summarize here the key conclusions we can draw from history.

First, the telephone was an example of "technology-push," users did not really know what to make of it and many may be surprised by what we take today as self-evident: obvious uses of the telephone were not at all so to early users.

Second, nevertheless, users adopted the technology and eventually even came up with uses not imagined by its inventors.

Third, the notion of the telephone being an analog to a voice conversation represented a very different, even unnatural principle at the time. The dominant concept was the telegraph and initially the telephone was conceived as voice telegraphy rather than as a separate and unique mode of communications.

Fourth, as data communications began to grow, new protocols and approaches emerged that were very different from those used for voice. Today data protocols dominate both voice and increasingly, television service networks.

Finally, despite the fact that content and video seem to be novel concepts, both figured prominently in the initial ideas of the inventors of telecommunications.

*

At the dawn of the second life of networks, it may be useful, even surprising, to recall the paradigms that have marked their first life. When Alexander Graham Bell invented the telephone in 1876 his objective was to enable people to communicate remotely, in real time, person-to-person ("point to point"), and primarily by means of speech. It was this last characteristic, the ability to transmit the spoken word, that made the telephone a major leap into modernity.

Admittedly, it was already possible to communicate at distance and point-to-point, thanks to the optical telegraph, then later to the electric telegraph. The latter, invented by Samuel Morse in 1832, was considered an important step in the development of telecommunications. In 1870, noting the rapid growth of telegraph messages, William Orton, then Chairman of Western Union, told Congress that the telegraph had become the "nervous system" of commerce[1].

So convinced was Western Union of the telegraph's rosy future that it refused to buy rights to use the Bell patent, available for $100,000 at the time. They found themselves forced a few years later to reconsider their position by offering $25 million to Alexander Graham Bell for the same rights who in turn refused.

However, the various forms of telegraph invented hitherto remained confined to the transmission of a coded signal that was then transcribed into text. The real breakthrough introduced by Bell and other inventors of his era (like Johann Philipp Reis, Thomas Edison, Charles Bourseul, and Elisha Gray) was the ability to transmit sound, and therefore the human voice. It is only when a mode of telegraph transmission based on a series of electrical impulses evolved to one based on the modulation of electric frequencies that the idea appeared to transmit electrical vibration and thus voice on the first telegraph networks. This represented the first step in "acoustic" telegraphy. The challenge was then to "talk" with electricity[2].

Alexandre Graham Bell

Source: Early Office Museum. (http://fr.wikipedia.org/wiki/Image:1876_Bell_Speaking_into_Telephone.jpg)

Bell, a specialist in elocution and teaching the deaf to speak, was Professor of Vocal Physiology and Elocution at the School of Oratory at Boston University[3]. This intimate connection between Bell and acoustics was finally recognized by giving his name (or rather part of his surname) to the unit measuring the intensity of sound: the Bel (later renamed and quantified as the sub-unit "decibel" so as to be more practical for engineering application).

The commonplace act of calling someone today is the result of the resolution of a number of technological challenges regarding the efficiency, quality, and usability of telephone communications. Before 1878, subscribers were directly connected to each other by wires that were sometimes, as was the case in the United States, installed by users themselves. With the advent of the first telephone switch in 1878, it became possible to connect to any subscriber *via* human operators who would establish between each pair of users the requested physical connections. The telephone later became fully automated with deployment of electromechanical switching systems, which were key for the worldwide expansion of the telephone. This is because it then allowed it to "scale."

During the twentieth century, the telephone network or "circuit switching network" gradually replaced the early wire-line communications networks. For example, written telecommunications evolved from telegraph to fax and then to email made possible by modem-equipped microcomputers or PCs. Other networks later emerged that were able to carry data more effectively, first so-called "packet-switched" X.25 networks (Telenet and Tymnet in the U.S.), and then the Internet where packets were transmitted as a continuous flow and consisted of discrete single data.

The Inherent Primitivism of First-Life Networks

The newly emerging telephone network, even in its early days, appeared as a much more complex and sophisticated network than others that existed at the time. For example, unlike electrical and hydraulic networks based on "point-to-multipoint" architectures where each user is connected to a centralized and common resource like a power plant or water tank, the telephone network is based on a "point-to-point" architecture where each individual subscriber can connect to another subscriber of their choice.

Another major technological feature was the fact that the telephone allowed "natural" conversations during a call using a method called "full-duplex." In other words, each of the two subscribers could intervene in the conversation (this is the "bilateral" nature of the link) and even interrupt each other (this is the "simultaneous" nature of the link). This type of connection can be compared, in particular, to the most basic "simplex" connection (where the signal is transmitted in only one direction, as is the case for connections between mouse and PC or PC and printer). Yet another method is "half-duplex" or "alternate" connection (where the signal can be transmitted in one direction or the other, but not both simultaneously) as is the case with the walkie-talkie.

However, during their first life, telecommunication networks were in many respects primitive and inefficient. In particular, once initiated, the electrical circuit carrying the call was dedicated and could not be used for anything else, even when the callers were not speaking. This is why, in the first life of networks, telecommunications services pricing was closely dependent on call duration and the distance between the subscribers. Conversations naturally include periods of silence that result from tacit agreements designed to avoid an unintelligible cacophony (it is only during extended silences or breaks that one of the parties to the conversation can verify that the link is still active by asking, "Can you hear me?"). These moments of silence, during which the telephone connection nevertheless remains "engaged" are ultimately wasted capacity. While this technical approach did admittedly result in a guaranteed quality of service, it also, in retrospective, introduced inefficiency in the use of expensive infrastructures.

In this context, it is interesting to remember that the phone itself was indirectly invented when scientists—including Bell—were trying to optimize the use of existing infrastructures for the telegraph. They were exploring ways to move from the principle of "one telegraphic conversation = one wire mobilized" to a more economical principle of "one telegraphic communication = one electrical frequency," with each wire being able to carry electrically several frequencies. The invention of the telephone thus evolved the network infrastructure from one type of inefficiency to another.

It was not until the development of information technology and electronics, then of the X.25, Asynchronous Transfer Mode (ATM) and Internet (IP) protocols that it became possible to transmit information by electronic "packets," therefore optimizing the use of the infrastructures (because no "packets" were delivered during periods of silence—see below). These protocols allowed the creation and sharing of rich information in the form of text, graphics, sound, pictures, and videos.

However, the efficiencies introduced by packet-based network protocols came at a cost, a trade-off. Networks lost their "state-fullness." This term refers to the holistic awareness by the network of the state of a given communication throughout its length. This awareness comes primarily through the maintenance of a "live" circuit throughout a telephonic conversation. One of the characteristics of the Internet Protocol is that state is not maintained. Each packet transmission is a separate and whole event in of itself making any kind of traditional guarantee of service quality untenable. We will see below how this lack of state manifested itself in other parts of Internet as well as how networks in their second life are attempting to regain "state-fullness" and even evolving the concept in social networks.

In terms of usage, the essentially bilateral and technically rigid attributes of telecommunications during the first life of networks actually relegated users to the role of passive and "cold" network infrastructure endpoints, and for a long time they could only be connected to one another *via* human beings ("operators").

An Unexpected Route to the Future

Although they may appear as out-dated today, first life networks, nevertheless, carried in themselves the seeds of their second life. For example, one of the key purposes of telecom networks, to carry and distribute "content," was implicitly embedded in the early days of telephony, twenty-five years before the advent of radio and fifty years before that of television.

Television's early history is still part of popular memory: the first regular broadcasts in the United States and Europe in the 1930's; the beginnings of Eurovision with the coronation of Queen Elizabeth II of Great Britain in June 1953; color television in 1951; the 1968 Olympic Games in Grenoble, France broadcast live to more than 600 million viewers in 32 countries; and finally, the first steps of the first human being on the moon, with the live broadcast of Neil Armstrong on July 21, 1969.

Long before 1940, at the dawn of the telephone's invention, the idea of tele-vision or vision at "distance" (from the Greek *tele* = far), had already taken shape. If it were possible to transmit sounds across an electrical carrier, it should indeed be possible to transmit images the same way. Bell was erroneously attributed the invention of the "telectroscope," in 1878. It was supposed to be an apparatus able to transmit images remotely and was an ill-fated ancestor to the television we know today. This seemingly improbable device relied on the wave nature of light and consisted of a multitude of wires transmitting simultaneously with utmost precision each color frequency. This bundle of wires would be woven into a cable that would carry the signal aboveground, under-ground, as well as underwater between continents. Bell did however invent with his assistant Sarah Orr an apparatus he called the "photophone" in 1880 which would encode and transmit sound over light waves[4]. Interestingly, this last discovery, which unfortunately never saw the light of day, would have allowed the creation of an optical telephone system effectively skipping the stage of telephone networks based on copper lines!

The rumor about a hypothetical device invented by Bell that would allow sight at a distance stimulated interest and research on the topic and, in 1880, in an article titled "Seeing by Electricity,"[5] by an electrician in New York City named W. E. Sawyer, the demonstration of a device that "sees at a distance through a wire telegraph" is mentioned.

In fact, the word "television" seems to have been used for the first time in Paris in 1900, but it was not until the 1930's that important steps in its development would be taken in the laboratories of the Bell Telephone Company. In 1927, it would be the first to demonstrate a live broadcast between New York City and Washington.

Beyond the technological breakthroughs that this "vision at distance"—the forerunner of television—would encourage, it is interesting to look back for a moment to the end of the nineteenth century and consider the remarkable services it was thought to make possible. Five major breakthroughs were anticipated: merchants would be able to display their goods anywhere in the world; police could quickly report escaped criminals to their colleagues in other countries; individuals would be able to remain in constant contact with their loved ones and, in combination with the telephone, to converse with them simultaneously; painters and sculptors could show their works anywhere and performing artists could remotely broadcast performances; and readers would be free to browse books in distant libraries or to transmit handwritten documents of their own to others[6].

Even though the ideas of broadcasting and interactivity were mentioned early in networks' first life, and to a certain extent were enabled first by photography, then television, it was not until the advent of the Internet that interactivity in the complete sense of the word would be available.

If the dissemination of electronic visual content became possible quite a while after the telephone, the dissemination of "audio" content was possible quite soon after its invention. For example, some entrepreneurs in the United States began offering very early on vocal-based content such as news, religious services, and weather reports as well as the announcement of sales. The telephone companies themselves also offered sports results, train timetables, and exact time of day services.

Another early application developed for the telephone was the "théâtrophone," invented by Clement Ader in 1881. It distributed music and live concerts, which were, until the advent of radio,[7] the first form of electronic distribution of culture and entertainment.

Another invention that complemented the telephone and also marked the immediate interest of scientists in the distribution of content over networks was the "phonograph" invented in 1878 by Thomas Edison. Edison actually sought to improve the use of the telegraph by inventing a recording system on a physical "memory" for messages pre-translated into telegraphic code. These messages could then be sent later, re-sent repeatedly or stored in a form other than paper. He therefore invented the sound recording

Ad for the Theatrophone

Source: Chéret J., Paris, Imprimerie Chaix, 1896
(http://histv2.free.fr/theatrophone/proust1.htm)

machine and its "solid" state memory specifically in the form of scrolls covered with thin sheets of tin that later was replaced by a wax substrate.

Contrary to its original purpose, the phonograph has almost always been used in "disconnected" mode from the network to listen to music, songs, and speeches, which became the dominant type of recorded content made available industrially first on vinyl and then on digital CDs. Edison believed, however, that his invention, combined with the telephone, would create new uses. He proposed ten in the *North American Review* in 1878:

1. Letter writing and all kinds of dictation without the aid of a stenographer.

Ad for Edison's phonograph:
"I want a phonograph in every home."

Source: Library of Congress.
(http://www.americaslibrary.gov/cgi-bin/page.cgi/aa/scientists/edison/phonograph_2)

2. Phonographic books for the blind.

3. The teaching of elocution.

4. Reproduction of music.

5. The "Family Record"—a registry of sayings, reminiscences, etc., by members of a family in their own voices, including the last words of dying persons.

6. Music-boxes and toys.

7. Clocks that would intelligibly announce the time (for going home from work, eating meals, and so on).

8. The preservation of languages by exact reproduction of dialects and pronunciations.

9. Educational purposes, such as the recording of lectures of teachers for later reference by students as well as spelling and other exercises intended to aid in memorization.

Finally, an example of early convergence thinking:

10. Connection with the telephone, making it a part of a solution capable of recording and transmitting permanent and invaluable records, instead of just receiving momentary and fleeting communication.[8]

The man who "[wanted] a phonograph in every home" the very homes in which 115 years later Microsoft's Bill Gates would predict an equally widespread adoption of computers, had in fact described, well ahead of his time, the answering machine.

Content, in text form, is for the most part even more closely linked to the telephone, which was developed to transmit voice over networks designed to carry text and numbers (the telegraph network). Yet it was not until the 1980's that transmission of text messages began to take-off, first with paging (through a paging device), then on-line messaging and chat-lines, and finally, with the quite unexpected and spectacular success of Short Message Service or "SMS" in mobile telephony in the 1990's.

More broadly, the phone quickly came to be recognized as a powerful tool for the dissemination of entertainment-related content. In 1877, in St. Louis, Missouri, a popular song The Wondrous Telephone, already imagined the type of entertainment enabled by having a telephone at home:

You stay at home and listen
To the lecture in the hall
Or hear the strains of music
From a fashionable ball!

Similarly, "voice" could arguably be regarded as the ancestor of today's user generated photo, music, video, and text content.

Beyond the original purpose of telecommunications networks in distributing content, there are some other "deja-vu" moments in the first life of networks that are interesting to recall.

For example, telephone switchboard operators acted in the early days of the telephone much like today's Internet search engines. Their role, which was to interconnect subscribers by linking the caller's circuit to that of the called person's by hand, was perceived as critical at the time for the smooth operation of the telephone system. There is an early story out of Lowell, Massachusetts that aptly demonstrates this point: "During an epidemic of measles, Dr. Moses Greeley Parker feared that Lowell's four operators might succumb and bring about a paralysis of telephone service. He recommended the use of numbers for calling Lowell's more than 200 subscribers so that substitute operators might be more easily trained in the event of such an emergency. The telephone management at Lowell feared that the public would take the assignment of numbers as an indignity but the telephone users saw the practical value of the change immediately and it went into effect with no stir whatsoever. (although attempts had been made, the implementation of dial telephone systems had yet to occur)"[9]. The fear of decimation in the ranks of telephone operators in Lowell, Massachusetts by a measles outbreak may have been the precursor to the telephone numbering system, which was not to arrive until much later[10].

The original telephone assistance service provided by human operators in the early days of the telephone may have also made the system subject to human foibles. Consider, for example, what is said to have happened to Almon B. Strowger, an entrepreneur undertaker in Kansas City, Missouri, who, after having discovered that his main competitor's wife, a telephone operator herself, was systematically diverting calls coming to his business to her husband's, undertook to invent an automatic telephone system that would operate without human operators[11].

One of the original purposes of telecom networks, echoed and amplified in their second life is the building of social networks among geographically distant individuals[12]. Indeed, phone networks were very quickly adopted by users and have become permanent parts of their private lives. This has facilitated the development of social relationship networks between and among urban centers and rural areas. Until the start of the Second World War, most of the residential subscribers in the United States had common lines, called "party lines," which were shared with their neighbors. While now very rare, they have not completely disappeared in the United States. The ring tones of "party lines" were different depending on which of the subscribers was called and all parties were encouraged to limit their time on the line. With such a system, it was common that subscribers simply picked up the phone when it rang, either by error—or in order to eavesdrop.

"Though the lines lacked privacy, they helped build a sense of community. Several calls in succession to the same number sparked worries that something was wrong, others would pick up and listen in to find out whether there was anything they could do to help. 'It wasn't really nosiness; it was neighborliness,' Helen Musselman of Hamilton County, Indiana, told an oral history interviewer in the 1980's. 'Now, she said, it's cold... You don't know what your next-door neighbor is doing' [she concluded]"[13].

1975–1995: The Long Transition

However, despite these initial similarities with what would become the second life of networks, for nearly a century the efforts of the industry were to be focused on the expansion and democratization of the services associated with first-life telecommunications networks.

Indeed, the figures speak for themselves. At the time of the opening of the first commercial telephone exchange in New Haven, Connecticut in 1878, there were 21 subscribers. Two years later, there were already 47,900 subscribers in the United States, 1.4 million in 1900, and nearly 81 million by 1963. In most other countries, the early development of the telephone was not as rapid. In France, for example, there were only 1.4 million lines in 1950, which represented a very low density relative to the US. It was not until 1980 that lines in France reached the 16 million mark[14].

Nevertheless, generally speaking, from 1876—the year of the telephone's invention—to the mid 1970's, which encompassed the transition to digital, the telecommunications industry has mainly focused on the gradual extension of service to virtually everyone in the world within range of the infrastructure. Because of this absolute priority, the period was characterized by the development of radio communications, the installation of submarine cables (first trials in 1927, followed by deployment of transatlantic cables from 1955), and the launch of communication satellites (initially in 1962, with Telstar 1). One of the major challenges faced was the interconnection of networks across the world. This led to the creation of sophisticated systems called international "interconnection" switches (not fully compatible in the early days), an international numbering plan, and an international telephone information inquiry service, as well as the basic rules for sharing revenues among operators. To achieve this, an international organization was created in order to organize the flow of global communications and establish *de jure* standards for the entire telecommunications sector: the International Telecommunication Union (ITU) based in Geneva, which in 1932 succeeded the International Telegraph Union founded in 1865 and in 1947 became a specialized agency of the United Nations[15].

During the First and Second World Wars, the deployment of civilian telecommunication networks slowed and was offset by the progress made in military networks largely due to their growing strategic importance in light of the military conflicts of the era. In the United States for example, Bell Labs, the R&D arm of the Bell System focused at the time, among other subjects, on research for the Defense Department. Research during this era greatly influenced network design principles. The roots of the Internet for example, grew out of a network design for the US Defense Department that would ensure the survival and resilience of civilian telecommunications networks in the event of a nuclear conflict[16].

When the United States entered into war in 1917, all developments in the field of radio[17] were under the control of the Navy in order to avoid any risk of espionage.

Despite the focus on expanding coverage and reach, first life telecommunications networks did, however, benefit from significant technological innovations, albeit ones coming from related fields. These technologies emerged from other types of networks being deployed at the time. Radio broadcasts to the public began in the early 1920's and commercial television in 1928. The electromagnetic coils used in the first telephones gave way in radios and televisions to electronic vacuum tubes, which began to be widely manufactured in the 1920's.

Furthermore, the invention of the transistor in 1947 and the development of the semiconductor industry were to play a decisive role in the evolution of computing and microelectronics, and more generally, in the development of related technologies. The transistor gave birth in particular to the solid state electronic memories and microprocessors that are found in all consumer and professional electronics equipment, including mobile phones, personal computers, and professional workstations such as those used in Computer-aided design (CAD) and Computer-aided manufacturing (CAM).

During this period, many of the telecommunications specific innovations proved to be too far ahead of their time to survive. These include: the transmission of images by telephone in 1924; the transmission of color in 1927; the coaxial cable (designed originally to avoid interference from electromagnetic waves); lasers (which would later re-invent transmission by use of optical fibers); and the video-telephone, which was developed in 1964, but still had to meet with lasting success. Today, web cameras coupled with Internet-connected PCs have become a viable and widespread substitute for video-telephones. By the 1970's telecommunications networks throughout the world were poised to go through a period of unprecedented technological transition marked by three "Big Bangs."

The First Big Bang: Digitization

The first Big Bang propelled telecommunication networks into a digital universe. All forms of information switched and transported by these networks were now represented as a series of electronic zeros and ones called "bits" (a contraction of "binary digits"). This Big Bang was in no way intrinsic to telecommunications: it was the result of developments in the adjacent fields of information technology and microelectronics. Both of these fields were themselves stimulated by the tidal wave of silicon and the exponential progress of integrated circuits that used silicon as their basis.

Since the early 1970's, the speed of innovation in integrated circuits has been dazzling as described by Gordon Moore's now famous law regarding transistors. In fact, there are several versions of this law. The first appeared in 1965 and asserted that the complexity of semiconductors doubles every eighteen months. Ten years later in 1975, Moore himself amended his law to state that "the number of transistors that can be inexpensively placed on an integrated circuit is increasing exponentially, doubling approximately every two years". This latter version (the one that is usually quoted) has yet to be contradicted. Even allowing for a slight downward revision, Moore's law faithfully reflects the rapid expansion of digital technology and its powerful impact on telecommunications networks.

Later, as telecommunications networks gradually merged with the world of computers, information technology itself became "communicating." Information technology began to insinuate itself in homes and offices throughout the economy and was now able to link computers by means of special networks having a wide and heterogeneous range of standards and protocols. This resulted from an unpredictable combination of commercial standards, some *de jure*, others *de facto*, including internet protocols as well as the "industry standard" established by Microsoft's Windows.

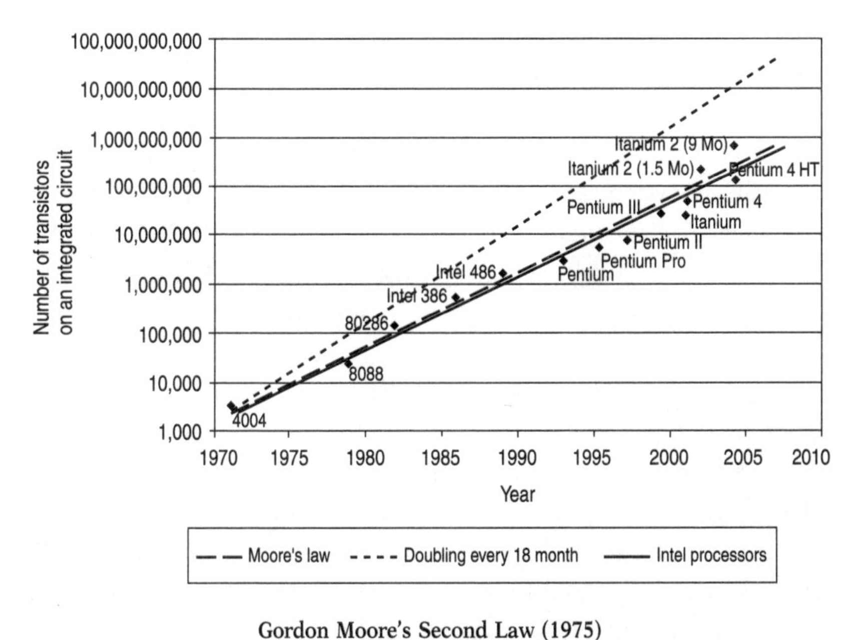

Gordon Moore's Second Law (1975)

Source: http://en.wikipedia.org/wiki/Image:Moores_law.svg

In the United States the first automatic telephone switching system (the ESS1 developed by Western Electric, which included digital components), was put into operation by Bell Labs in 1965. Although the system still deployed some electromechanical relays, its use of computers represented a significant advance over the fully electromechanical automatic telephone switches of the day. Their key virtue was that they could be easily programmed and customized. These hybrid systems were in turn quickly supplanted by the fully digital systems that appeared in the 1970's (like the E10 in France). These were based on the vocal sampling of conversations and their subsequent trans-coding into bits of computerized information (as in the case of music CDs). This digitized information could then be routed to its destination *via* a fully computerized system of switches and transmission networks that could combine many simultaneous digitized conversations into a single physical channel. However, the connection of telephone subscribers to the network still relied on traditional "analog" access lines that carried electrical signals (like those of a vinyl record) to the central office where the telephone switch established and digitized the signal.

This wave of digitization gradually spread throughout the telecommunications system, from the telephone exchange to the user, *via* the deployment of "integrated services digital networks" (ISDN), designed from the outset to deliver high-speed data connectivity from the subscriber to the local network access point without relying on a low speed "modem"[18]. Such networks were designed to meet the growing need of businesses to exchange rapidly large computer files directly, which up to now could only be done through traditional postal service or courier (often referred to as "snail mail" in view of the delays to which it gave rise). In addition to their primary function of transferring computerized messages such as data files and email, these ISDN networks were also used to improve significantly the quality of vocal communications by means of an end-to-end digital format.

In some ways the digital revolution represented a return to basics. The analog model of first-life networks, according to which an electrical signal is modulated to match the undulation of a voice, was itself a departure from the way our ancestors communicated at distance. In ancient times human beings used "discrete" (i.e., well differentiated) signals to write a finite number of letters signs and words, or fingers to count—thus the word "digital" (from the Latin word "digitus" for finger). The earliest means of communication were themselves also based on simple and discrete signals (like smoke signals, the telegraph or flag semaphore). In ancient Greece, the legend of Theseus demonstrated the use of binary signal communication—and the related risks (in the absence of an "error correction protocol") that we are familiar with today. On his return to Athens after killing the Minotaur[19], Theseus forgot to use the binary code of communication established with his father: hoisting white sails if successful, black in case of failure. Convinced of the death of his son upon seeing an inadvertently hoisted black sail, Aegeus committed suicide by drowning in the sea that now bears his name.

The legend of Theseus

Source: http://www.dl.ket.org/latin1/gallery/myth/heroes/theseus.htm

Beyond the new ability to communicate with the world of computers, the advent of digital technologies in telecommunications networks in the late 1970's brought many technical and economic benefits. First, the trans-coding of vocal samples as well as of textual characters, pictures and videos into discrete elements independent of each other introduced by its very nature a new flexibility in telecommunications networks. Digitization itself gave rise to new and more flexible modes of transmission over existing networks characterized by new protocols for delivering packets of bits (or "datagrams") that could be stored and moved at will, a great advance over the old method of analog signal transmission (see below.) In addition, the coding of information based on "zeros" and "ones" significantly reduced the probability of error. It was easier to distinguish and if necessary regenerate discrete states (zeros and ones) than to faithfully reproduce a continuous curve (an analog signal of voice, including successive minima and maxima of different scales).

These technical improvements and advantages in signal transmission came in addition to those that developed for computers. In particular, the techniques developed to allow simple information coding and sequencing could be used to handle, copy, store and preserve information almost indefinitely, heralding the shift from vinyl to digital CDs and from videotape to DVDs.

From an economic point of view, the digitization of networks substantially reduced costs, both in terms of investment and operations. This became possible with the development of the microelectronics industry and the mass production of miniaturized components and products. This miniaturization also resulted in a reduction of the physical footprint occupied by digital telephone switches, without adversely affecting either their speed or their ability to provide new services to subscribers. Installation time was shorter as well, and the absence of moving parts improved reliability and reduced the cost of maintenance. Nevertheless, the high concentration of electronic components increased the cost of cooling because of greater heat dissipation.

The Second Big Bang: The Internet

The second Big Bang that characterized this period of transition resulted from a complex assembly of technologies that led to the development of the Internet. On the one hand, companies had started to deploy dedicated computer networks for their internal operations that used specific transmission protocols i.e., a set of rules for communication, such as the Ethernet protocol, developed at the Xerox Parc research center in Palo Alto, California between 1973 and 1975. On the other hand, telecom operators in most developed countries began in the late 1970's to invest in public data transmission networks for the purpose of connecting remote computers (such as Telenet and Tymnet in the US, and Transpac in France, based on the X.25 standard in 1976). These networks, which themselves were interconnected to existing telephone networks in order to reach a large established base of users, were designed to operate "on top of" these networks and to transmit data (not voice).

The effects of the mass production of Ethernet based equipment (led by new companies like 3Com, Cisco, Intel and Digital Equipment Corporation) led to the spread into all data networks of the technical standards that were to define and become synonymous with the "Internet." The Internet consequently began in early 1983 its rapid worldwide growth.

The Internet, or rather its ancestor "ARPAnet," had been designed as early as 1967. The creation of the DARPA [Defense Advanced Research Projects Agency] in February 1958 was directly attributed to the darkness of the Cold War and to the launching of Sputnik on October 4, 1957. The U.S. government realized then that the Soviet Union had developed the capacity to rapidly exploit military technology.[20] The characteristics of this early prototype deserve to be mentioned here because they were to have a lasting influence on networks and shaped in part their transition into their second life.

First, unlike telephone networks, which used a hierarchical arrangement of switched circuits, the Internet is "flat" (i.e., fully distributed and decentralized), so that individual components can operate independently of others—an essential source of resiliency in the event of breakdown or the destruction of a part of the network.

The switching technology associated with this flat network, known as "packet-switching," was made possible by the digitization of information. In this model, computers simply parse or sequence messages (voice, data of all kinds) to be sent into small chunks of varying sizes called "packets" or "datagrams."[21] Each datagram is an independent entity, like a letter sent by regular post, which includes information about the sender and recipient, as well as the information needed to reconstruct the entire message[22]. The various components of the message can therefore be sent *via* different paths to their destination in the event that "traffic jams" occur in some parts of the network during the course of transmission—in the same way that the numbered pages of a book could be sent to the same destination *via* different post offices, since the page numbering allows the book to be perfectly reconstituted in its original form upon arrival. Similarly, each "leg" of the network carries a succession of datagrams that have, for the most part, nothing in common with each other i.e., they come from different and independent messages carried on the network from different conversations.

The huge advantage of this technology over networks using switching circuits consisted in its elimination of problematic "silent time" and in the optimization of the resulting infrastructure. Indeed, it is no longer necessary to engage a dedicated channel throughout the entire transmission session.

Another key feature of the Internet is the idea of "best effort" delivery. Unlike the telephone network, in which the mobilization of the physical line for the entire communication ensures the quality of service (i.e., the full delivery, end-to-end, of the voice conversation), the Internet provides no guarantee for delivering the information to the addressee (quality of service), nor even a system of priority management. The outcome depends in practice on the network load at any given moment. This is why in some cases information may be lost. The loss of a packet is seen in the case of a pixelized Internet-connected television screen when individual pixels take on a strange color, contrasting with those of their immediate neighbors.

This model of "best effort" can be applied in other adjacent Internet contexts. In these other cases it resembles a tacit mode akin to the principle of "good enough." In the new models of information and communities operating on the Internet, individuals can for example, provide information, advice, opinions, and recommendations without a "higher authority" intervening to ensure and guarantee the quality and reputation of the source. The growing abundance of such sources leads to diversity, complementarities, and redundancy, which act to offset this congenital weakness in reputation assurance.

The Third Big Bang: Mobile Networks

In the early 1980's, telecom operators began investing in mobile networks—the third and final source of momentum for the development of the second life of networks. The decrease in costs resulting from economies of scale in the production of silicon and critical electronic components is again central in this global development. Above all, the very early adoption in 1987 of a common standard for mobile communications called GSM (Groupe Spécial Mobile initially, then renamed Global System for Mobile), made possible the large-scale production of terminals and network in the wireless sector.

Radio communications began to be developed before the Second World War, when they were used to support communications between military and police vehicles. These "point-to-point" communication systems were based originally on alternate (rather than full-duplex) exchanges: callers needed to take turns speaking and listening, which led to the exclamation "Over!" used to signal the end of a phrase requiring a response by the other party. These first analog mobile telephone networks were not connected to the traditional landline telephone network. Additionally, the equipment, and terminals used in these first-generation transmitters were heavy, bulky, and energy-intensive, and signals could only be transmitted over a limited geographical area.

Research in mobile communications started in 1930 led, somewhat indirectly and haphazardly, to discovery of the modern mobile phone in 1947. The first key idea was to use low-power transmitters to cover small areas (later called "cells") with a signal. This allowed reuse of frequencies (always in limited supply) in nearby areas without causing interference. Each cell is associated with a two-way radio called a "base station," so that the transition of a mobile telephone from one cell to another occurs without interrupting the connection. This seamless "hand-over" of a session is the key technical strength of the system. The ideal configuration of such a network was based on hexagonal cells (as in the structure of beehives), which interlocked more economically than circles giving full coverage. A December 1947 Bell Labs technical memorandum[23] describes that they had in fact originally considered three geometric shapes: triangle, square, and hexagon. These three forms share a crucial property, namely that the distance between the center of the cell and its maximum perimeter is uniform. Of these three forms, however, the hexagon proved to be the most effective, maximizing the number of mobile phones that can be reached from a given base station. The actually constructed cells were not perfect hexagons but looked more like a kind of amoeba. In addition, they overlapped each other in order to make it possible to increase capacity in response to growing demand and to leave no area insufficiently covered. From a service point of view, cellular systems allowed the availability of the service to a potentially very large number of subscribers, allowing them to move freely and quickly without a loss of either the quality or the call itself. The

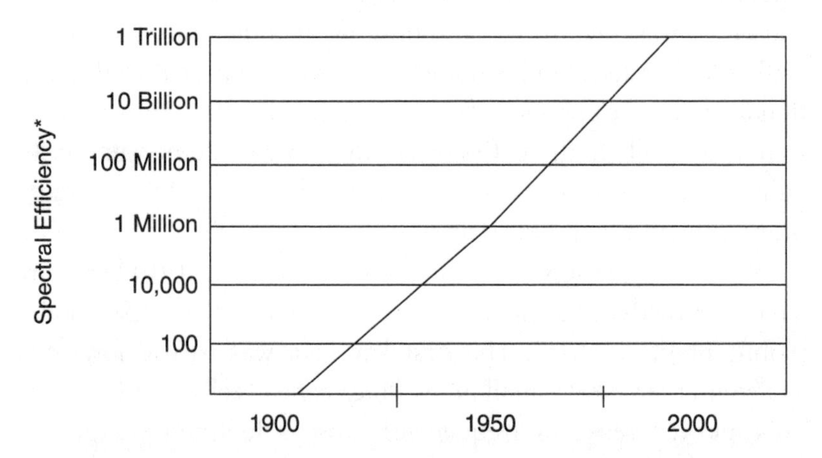

Cooper's law

Source: http://www.arraycomm.com/serve.php?page=Cooper

sound quality of such communications was actually quite close to that of fixed line telephones.

The first demonstration of a prototype cellular phone was on April 3, 1973, by an employee of Motorola in the streets of New York, a certain Martin Cooper[24]... the very man who stated the law bearing his name and that states that the number of radio frequency conversations which can be concurrently conducted in a given area doubles every 30 months.

The analog mobile phone quickly raised security and confidentiality issues. Anyone with a radio scanner could listen in to the communications of others and even identify their phone numbers, allowing eavesdroppers to make fraudulent calls. Moreover, it was impossible to use such mobile phones in different countries. Analog mobile phones were bulky as well and so not truly mobile: in the 1970s, they were found mainly in cars and came to be known as "car phones." It was the limited number of possible communication channels, however, that definitively doomed the long-term prospects of analog technology.

As was the case for fixed line telephones, the introduction of digital technology in mobile telephony would allow the development of a second generation of mobile telephones. This together with the adoption in Europe and in other countries of a common standard—the *sine qua non* for large-scale production, led to the drop in prices that occurred in the second half of the 1990's.

In 1982, the European Conference of Postal and Telecommunications Administrations (CEPT) established a "Groupe Spécial Mobile". This committee's goal was to develop a common European standard for digital cell phones. On September 8, 1987, twenty countries signed a technical proposal, which was then studied by the European Telecommunications Standards Institute (ETSI) between 1989 and 1991. At the same time, the first GSM network was launched in Finland in 1991.

The GSM standard provided for better voice quality and allowed additional services, such as the sending of short text messages

**The first commercial mobile telephone:
The Motorola dynaTAC TM 8000X (1983)**

Source: Motorola. (http://www.motorola.com/mot/doc/0/637_MotDoc.pdf)

(SMS). Furthermore, because transmission was digital, information could be encrypted ensuring greater confidentiality. An original feature of the GSM standard was the "SIM Card" (Subscriber Identity Module), a "smart" electronic removable card, which included the user's ID, telephone contacts, and preferences[25]. This in turn made it possible for users to use different mobile telephone handsets while keeping all of their information on their SIM card.

Implementing the GSM standard required overcoming cultural, industrial, political, and technical barriers in the form of differences between companies, users, regulators, and governments. This triumph is arguably one of the greatest technological achievements of European industrial cooperation. Things did not always turn out so well, as in the case of previous disagreements over standards governing color television and videotex. The remarkable success of GSM brought concrete benefits not just to Europe but to most countries in the world. For example, the mobile phone began to be widely adopted in the United States in 1995. Even though it coexists with a competing standard in the US—the CDMA standard—GSM remains the only service that allows both portability and use of the same mobile telephone in Europe and the United States. The figures demonstrate the massive impact that GSM has had in driving the worldwide adoption of mobile telephony. When the 1987 agreement was signed, its creators foresaw twenty million subscribers by the end of the second millennium; by 2000, however, there were already 250 million subscribers, and by the beginning of 2008 more than 2.6 billion GSM subscriptions, or about 85% of the global market in mobile telephony, corresponding to 1.6% of global GDP. Every year, mobile phone users are buying more than a billion new mobile handsets, generating over 7 trillion minutes of calls, and sending approximately 2.5 trillion Short Message Service (SMS) text messages.[26]

The comparative history of the penetration rate of mobile telephony in Europe and the United States demonstrates the positive role of standardization in the field of telecommunications. Whereas in 1996 the penetration rate of mobile phones amounted to nearly 20% in the United States, it did not exceed 10% in the European Union. Over the next ten years, the situation was almost reversed: the penetration rate in the European Union is now in the range of 100%, while in the United States it stands at about 70%[27]. Not only was the success of the GSM standard instrumental in the early development of mobile communications, but its philosophy of technical and international consistency continues to this day to serve as an example of how the coming new generation of mobile systems and phones can be effectively launched.

*

The 1990's were characterized by the availability of widespread digital telecommunications, Internet, and mobile networks. Yet many still believed that this revolution consisted only of gadgets meant for people who were primarily concerned with improving their social status. Few consumers yet grasped the intrinsic usefulness of these devices and services—rather like Proust's character who doubted the early promise of first-life networks: "It's very tempting, but rather in a friend's house than at home... Once the first excitement is over, it must be a real headache (*un vrai casse-tête*)."[28] In the middle of the 1990s, the disruptive technologies developed over the previous twenty years were facing similar skepticism. Indeed they remained inaccessible to the majority of people, due in part to the apparent difficulties in using them.

CHAPTER 2

The Gestation of Second Life Networks

By the mid-1990's complementary technical and economic trends in personal computers and mobile telephones allowed a majority of traditional fixed telecommunications users to adopt rapidly these new technologies.

According to McKinsey & Company[1], electricity reached one-quarter of Americans 46 years after its introduction. Telephones took 35 years and televisions 26 years. In just six years, Internet access had already reached 25 percent penetration.

The period bounded by the IPO's of Netscape and Google, which occurred in 1995 and 2004 respectively, defines the "gestation" period that preceded networks' re-birth. During this period, networks have developed two dominant traits: ubiquitous reach and permanent connectivity.

The Development of Information Technology: "Simple is Beautiful"

The rapid adoption of the new technologies formed by the successive "Big Bangs" of the 70's and 80's in telecommunications was driven first and foremost by the simplification of information technology tools that began to be available to the mass market in the

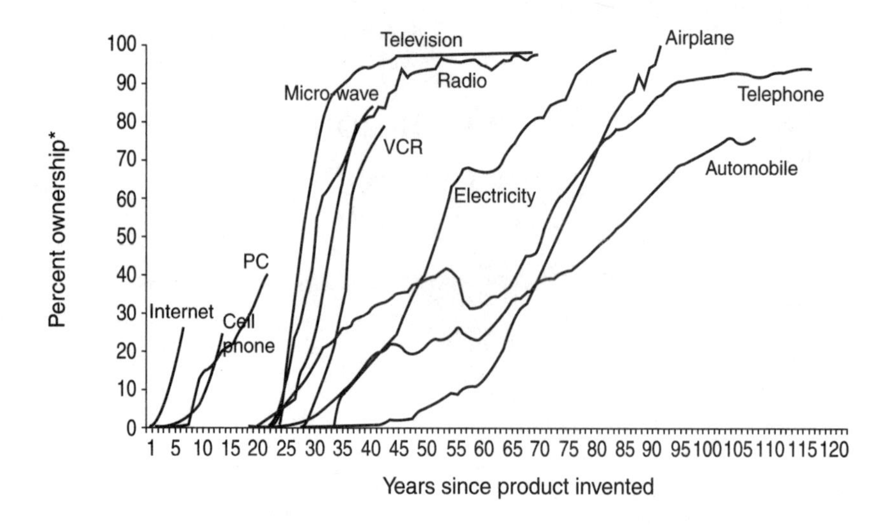

* Most of the time, "percent ownership" refers to the fraction of households that enjoy each product.

Sources : U.S. Bureau of the Census (1970 and various years);
Cellular Telecommunications Industry Association (1996); The World Almanac and Book of Facts (1997).

Adoption curves for various technologies

Source: http://www.planetd.org/2007/04/13/the-evolution-of-climate-change/

1970's. In fact, it is during this period that computers, hitherto available only to large enterprises and academic institutions, began to approach reasonable enough cost and size to attract the attention of consumers.

The arrival of the Apple II microcomputer in 1977 marks the beginning of the era of "simple is beautiful" information technology. Steve Wozniak and Steve Jobs, previously employed by Hewlett Packard and Atari respectively, conceived this revolutionary computer[2]. They developed a single player version of the game "Pong", called "Breakout", which used a smaller number of integrated circuits thereby reducing both the cost and complexity of its manufacturing[3]. To improve usability of the system, they added a keyboard (something already integrated in the Apple I launched in 1976 but which saw a limited production run) instead of the rows of buttons that were usually found in computer kits (like the Altair) and a video screen instead of the bulky and difficult to use printer.

The important influence of games on the early designs of "simplified" microcomputers cannot be ignored. This preoccupation with gaming led the industry to focus constantly on making user interfaces immersive and intuitive. This was a significant factor in adoption of the microcomputer by growing numbers of consumers.

Another factor has contributed to simplifying information technology and in particular computers. During the 1960's, when computers and their use began to be formally taught in certain American high schools and universities, students could only connect to these institutions' central computers under strict conditions. In order to allow the largest possible number of students to use computers, access to computing resources was limited to a small window of time. The small but growing group of those that had a special aptitude for computers and wanted to "tinker" and develop personal applications were hobbled by this Draconian system of rationed computer resources. Even after they entered the work force, they found similar constraints in the companies that employed them. This created an entire generation of students motivated to find a solution that would let them develop their own programs. This was instrumental in contributing to the development of individualized or "personal computing" as it would come to be called. The most well known member of this new generation is Bill Gates who started his first software project when he was a student at the Lakeside School in Seattle using a time-share computer through a primitive terminal with no screen.

The arrival on the scene of the first Intel microprocessor in 1969, the "4004," allowed a transition from the massive computers that filled entire rooms of enterprises, government, and universities—and consumed energy accordingly—to the more compact, less energy intensive microcomputer. A few years later, work done at SRI International together with the development of graphic interfaces by Xerox Parc researchers gave birth to the first "user-friendly" personal computer. This led to development of Xerox's

Alto Computer and Apple Computer's Lisa, some of the first examples of PC's with intuitive user interfaces.

The Internet "Toolbox"

Even though use of the Internet depended exclusively on existing voice telecommunications networks, in no way is it an extension of the telephone service. Telephone calls are initiated in two types of circumstances: (1) When we know precisely whom we want to call (in this case we dial the number after maybe having found it in a personal or public directory, or obtained it through directory assistance); or (2) When we have a well-defined need for a service (in this case we consult a professional directory or the "yellow pages" for one or several phone numbers of appropriate businesses).

The process is wholly different for the Internet. Its large number of available on-line services and content typically exceeds by several orders of magnitude the number of entries found in traditional "yellow pages" directories. At first, the familiar approaches learned in the first life of networks were re-used in the early Internet. For example, Yahoo! and other companies created directories or service guides that acted as sort of an "Internet yellow pages." There exists about 165 million Internet sites[4]; this would represent a paper directory the size of say, a 6 story brownstone. Pushing this analogy a little further: since each site has approximately 300 separate pages, we are now looking at a super-directory half the size of Mount Everest.

It became clear in the 1990's that beyond the progressive simplification of information technology over the previous twenty years, the growing adoption of the Internet required the development of a "tool-box." This would allow both technically advanced as well as not-so-advanced users to explore and discover this new universe of abundantly available, globally dispersed information services.

The primary tool in the toolbox was to be the search engine, a piece of software that allowed users to obtain a list of Internet sites whose content best fit the key-words entered by the user. The most popular search engine today is the one developed by Google.

In December 2007, it accounted for some 41 billion of the world's 66 billion Internet searches conducted for that month, or roughly 62% of the total. The other widely used search engines are by far behind this number: Yahoo! (8.5 billion searches or 13% of the total); Baidu (whose growing importance has earned it the label of the "Chinese Google" with 3.4 billion searches or 5.2% of the total); Live Search (Microsoft's search engine had 1.9 billion or 2.9% of total searches); Naver (search engine of the Korean NHN Group had 1.6 billion or 2.4% of total searches); and finally eBay, itself a search engine specialized for person-to-person auctions (with 1.4 billion or 2.2% of total searches)[5].

"I can't explain it—it's just a funny feeling that I'm being Googled."

Source: Charles Barsotti, *The New Yorker*, October 28 2002

Another major tool in the Internet toolbox, the "browser," appeared before the search engine and perhaps is more fundamental in that it represents the keystone of the Internet's architecture. The browser allowed two things. First, it provided a method to render Internet sites' content ergonomically and graphically. Second, it gave users a way to navigate easily, quickly, and logically across Internet sites. Quickly thereafter, graphical browser software started to appear on microcomputers. The original maritime metaphor of navigation has become even more apt with time. Today moving from one Internet site to another is fluid and effectively "seamless." An additional feature of browsers that became important over time is the way they simplified use of electronic mail ("email") including allowing users to access their mailboxes from any computer or web site.

Core to browsers is "hypertext" technology, invented by Ted Nelson as part of his Xanadu project in 1963 and Douglas Engelbart (inventor of the "mouse"), as part of his oN-Line System (NLS) project at SRI International. Hypertext permits the linking of certain words or phrases on a computer screen to documents that upon activation can take a user directly to complementary text either in the same document or in other documents resident on the computer. For example, hypertext allows a user to move smoothly and rapidly between text and footnotes as well as text and references at the end of the document. Hypertext, already widely used in information technology of the 1960's and 70's, represented a substantial leap in simplifying access to information and in streamlining the user-machine interface.

By the end of the 1980's, twenty years later, Tim Berners-Lee, a researcher working at CERN (Centre Européen pour la Recherche Nucléaire or the European Organization for Nuclear Research) in Geneva, Switzerland, succeeded in marrying hypertext, an essentially offline functionality, with the Internet, giving birth to online "hyperlink" technology.

Hyperlinks allow for the dynamic organization of information by incorporating in all electronic documents the potential to reach complementary documents through a simple click of a mouse on a word or phrase. These documents can be stored either locally on the computer used or on any server in the world connected to the Internet.

Hyperlinks represent one of the more decisive steps on the path to networks' second life. Having passed through the stage of networking telephones, the networking of computers came next, followed closely by the networking of websites and documents. By anal-

Declaration

The following CERN software is hereby put into the public domain:

- W 3 basic ("line-mode") client
- W 3 basic server
- W 3 library of common code.

CERN's intention in this is to further compatibility, common practices, and standards in networking and computer supported collaboration. This does not constitute a precedent to be applied to any other CERN copyright software.

CERN relinquishes all intellectual property rights to this code, both source and binary form and permission is granted for anyone to use, duplicate, modify and redistribute it.

CERN provides absolutely NO WARRANTY OF ANY KIND with respect to this software. The entire risk as to the quality and performance of this software is with the user. IN NO EVENT WILL CERN BE LIABLE TO ANYONE FOR ANY DAMAGES ARISING OUT THE USE OF THIS SOFTWARE, INCLUDING, WITHOUT LIMITATION, DAMAGES RESULTING FROM LOST DATA OR LOST PROFITS, OR FOR ANY SPECIAL, INCIDENTAL OR CONSEQUENTIAL DAMAGES.

Geneva, 30 April 1993

W. Hoogland
Director of Research

H. Weber
Director of Administration

ogy, the hyperlink concept is today finding relevance in the context of the relationship of individuals in social networks. These networks are described by the construct called the "social graph," perhaps not invented but certainly made mainstream in May 2007 by Facebook who described it as "the network of connections and relationships between people on the [Facebook] service." This is a very important component in the build-out of second life networks. As hyperlinks trace the relationship of documents and sites, the social graph describes the relationship between individuals. The number of sites that link to a site measures its importance. Similarly, the number of individuals that link to a given individual is a measure of their popularity and influence within the wider social network.

From the user's perspective, the Internet—which consisted of a worldwide physical network, built on first life telephone networks —fades into the background while the World Wide Web, the network of Internet sites and documents, occupies the foreground. In other words, hyperlinks are to the World Wide Web what copper telephone wires and the associated network transport protocols (IP, TCP) were to the Internet. Technically speaking, hyperlinks trigger actual physical connections between the displayed document on the user's browser and an information server situated somewhere else on the Internet network (a similar technique already appeared with the Minitel in the 1980s)[6].

Search engines began to leverage the underlying hyperlink structure that developed among Internet sites. An important part of these algorithms relies on the relevance of a site to a particular subject. In order to measure the relevance of a site, a proxy for this metric is used: the frequency with which a site's documents are referenced, that is to say, the number of web sites that contain hyperlinks pointing to the site whose relevance is being measured. The principle of searching the Internet is based on the very important tenet that the popularity of a site, as measured by user's "click-votes," is a strong correlate of its relevance. This relevance is contextualized to the specific keywords a user selects when initiating a search.

It was not until 1993 that the first functional and ergonomic Internet browser appeared. Mosaic, as it was called, was developed using the navigation technology ViolaWWW by two students at the NCSA (National Center for Supercomputer Applications), at the University of Illinois. Mosaic was initially used by universities and software companies[7].

The Netscape Navigator was the pioneering application most responsible for popularizing the use of the browser. It was created in 1994 by a team of ex-Mosaic researchers working at Netscape Communications Corporation. Netscape began selling its browser early in 1995, and saw in a few short months a spectacular uptake by users. The successful IPO later that year testifies to the exceptional nature

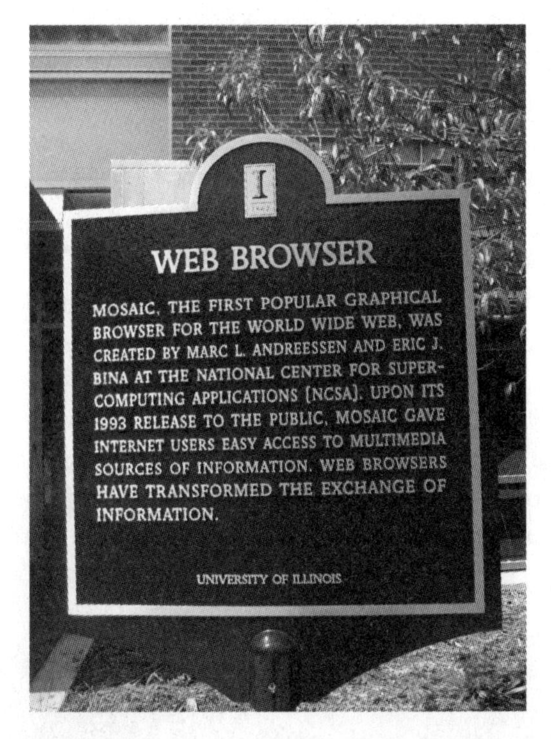

The plaque commemorating the creation of the Mosaic web browser at the University of Illinois

Source: http://en.wikipedia.org/wiki/Image:Mosaic_browser_plaque_ncsa.jpg

of Navigator and represents a tipping point in the mass adoption of the Internet. This is why, at least symbolically, 1995 represents the beginning of the "gestation" of networks' second life. The progressive adoption of the Internet by users would lead to a veritable amalgamation of Internet physical networks with the social networks created by its users that is the hallmark of second life networks.

A new generation of browsers subsequently appeared including Mozilla's Firefox developed by the open source movement, Safari developed by Apple, and Opera developed by the Norwegian company Opera Software[8].

Microsoft, having essentially ignored the Internet to this point, began seeing it as a threat to its leading product, Windows (even though Windows was an interface between the user and the computer, not between the network and the user). Microsoft decided to enter the race and announced the launch of its Internet Explorer browser (which it characterized as an "explorer") in August of 1995. Soon after that, it was packaged with Windows becoming available on virtually all Windows-based personal computers and included subsequently in all Windows updates.

This led to an inevitable "browser war" which ended three years later in 1998 with the purchase of Netscape by AOL (which has since then, itself become a part of Time Warner). This war was personified by the landmark antitrust case brought against Microsoft by the US Department of Justice (DOJ). The government argued that Microsoft was abusing its dominant position in personal computer operating systems by packaging Explorer with Windows.

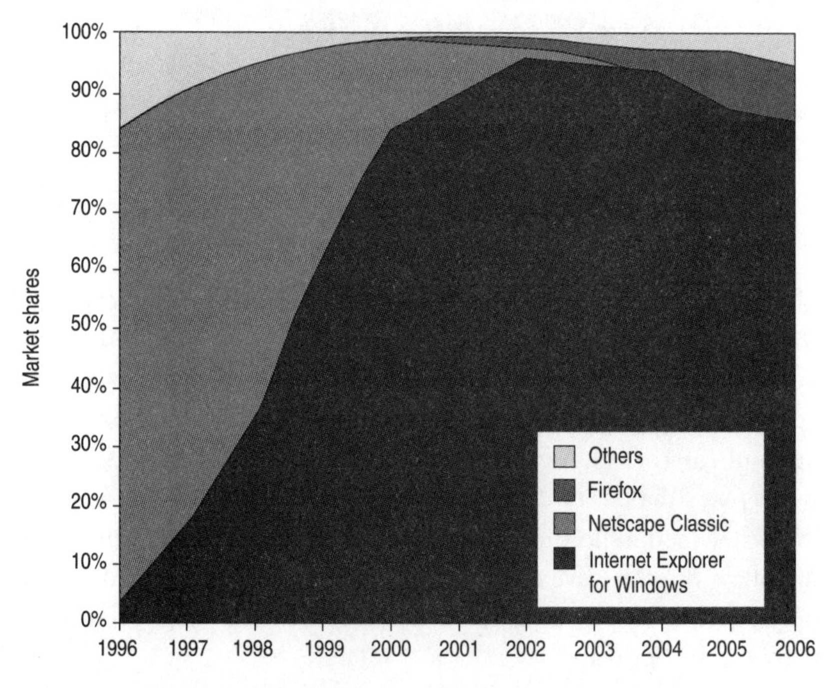

Market shares of rival browsers between 1996 and 2006

Source: http://en.wikipedia.org/wiki/Image:Browser_Wars.svg

Competitors such as Netscape were disadvantaged by the fact that users had to buy their software separately or download it over slow modem connections. Microsoft was accused of trying to monopolize the browser market through the practice of bundling. The District of Columbia Court of Appeals later overturned the original judgment against Microsoft. In the end, Microsoft settled by opening their Windows Application Programming Interfaces (API's) to third party companies.

The browser landscape today is dominated largely by Microsoft who held in November 2007 about 81%[9] of the worldwide market. Firefox, which came on the scene in November 2004 has already reached 13% market share, while Safari had 3% and Opera 1%, on a user base of more than 1.3 billion worldwide[10]. After 14 years of existence, in February 2008, Netscape closed its doors and encouraged its users to migrate to Firefox[11].

The emergence of a Windows integrated Internet Explorer has resulted in a *de facto* quasi-monopoly on the world's installed base of personal computers that has clearly contributed to the rapid adoption of the Internet throughout the world. The increase in the worldwide number of PC's capable of connecting to the Internet through a browser goes hand-in-hand with the increased use of the homogenous "WinTel"[12] design that leverages the sustained improvements in Intel microprocessors and the Windows OS. In any case, sustained competition between the two browsers led to a race to implement productive innovations, which ended with the rapid introduction of useful functionalities.

This rapid evolution, nevertheless, was firmly rooted in the way the Internet was used, despite the appearance of new types of communicating devices like portable telephones (including the new "smartphones" and multimedia portable phones) or PDA's[13] capable of connecting to the Internet, as well as the Internet consoles for televisions; Internet access remained the realm of the PC.

Despite the skepticism that emerged regarding the Internet in the decade between 1995 and 2004, the Internet has irreversibly changed the world. The estimated number of Internet users in the

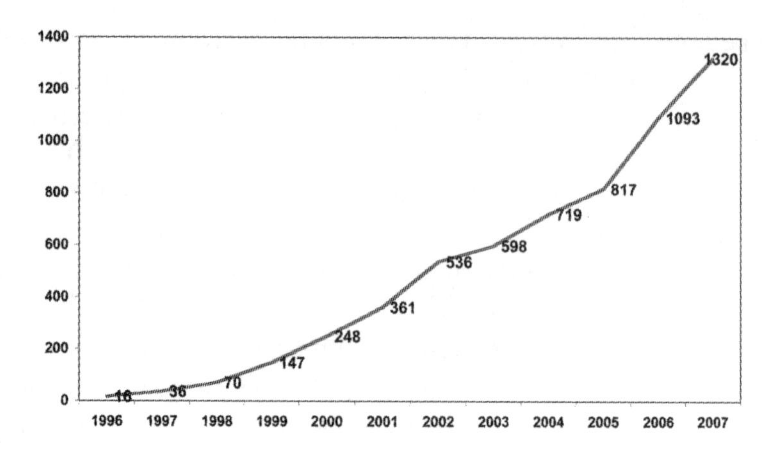

Growth in the Worldwide Number of Internet Users as of December of Each Year (in millions)

Source: http://www.internetworldstats.com

world on March 31, 2008 was 1,407 billion and the penetration 21%[14]. According to the monthly survey by Netcraft, the Internet grew from a mere 20,000 sites in August 1995 to more than 160 millions at the beginning of 2008[15].

Between 1996 and the end of 2001, traffic on the Internet backbone, for the United States, went from 1,500 terabytes per month to about 30,000 terabytes per month (1 tera-byte = 1,000 billion characters of all kind or 8,000 billion bits, each bit representing a zero or one—or 3 million copies of this book!)[16].

The Development of Broadband

In addition to the creation of tools that have permitted a wide range of users to access the Internet in the simplest possible way, the "gestation" period of the second life of networks was characterized by an acceleration in the speed of fixed access networks at a pace never seen before. In practice, any quantitative analysis of service usage requires that high-speed access penetration be factored in addition to factors such as the worldwide penetration of Internet equipment.

The early Internet was accessed by a modem connecting the computer to the telephone line that allowed an access speed that today would appear antediluvian. The maximum attainable access speed during the middle of the 1990's was 56 kilobits per second or 150 times less than what is possible today.

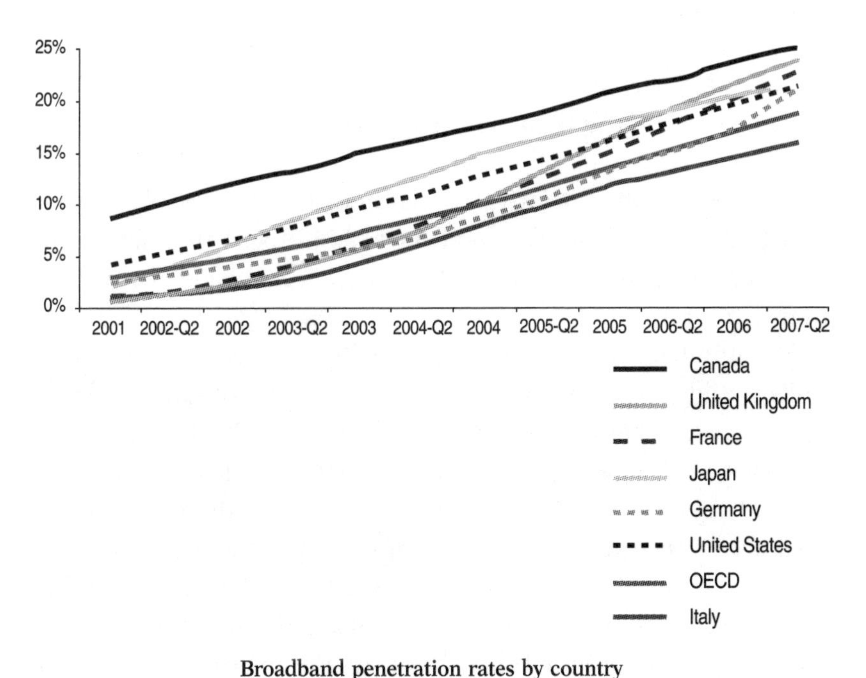

Broadband penetration rates by country

Source: OCDE.
(http://www.oecd.org/dataoecd/22/14/39574797.xls)

In Europe, the evolution to higher fixed access bandwidth speeds was achieved by adopting ADSL technology ("Asymmetric Digital Subscriber Line") almost exclusively, while in the US, due to the coexistence of telephone and cable companies, adoption of ADSL over telephone lines was complemented by the development of broadband over cable. For several years, in the US, only cable modem service could grant access speeds superior to 1 megabit per second. Cable modem service was originally marketed through two companies in the US, @home and Roadrunner (controlled by Time Warner). In 2007, cable remained the only way in the US to obtain access speeds of 6 megabits per second and higher. This is still slightly slower than observed speeds in Europe and certainly less than speeds available in Japan and South Korea where optical fiber is utilized in their access networks.

A study by the Pew Research Center showed that in 2000, five million Americans had broadband access. Just four years later, that number grew to 60 million and if enterprise access is included, to 72 million[17]. This type of adoption pattern is comparable to those experienced in the majority of developed countries. The most important aspect of the development of high-speed Internet access is its impact on the adoption rate of new technologies by a large number of users.

The numbers demonstrating this point are clear. According to the Pew study mentioned above, penetration rates of low-speed and high-speed Internet access approached parity in 2004. During this same period, US low-speed bandwidth users indicated that they spent 90 minutes per day on the Internet, while those with high-speed access reported spending 110 minutes per day. Likewise, 69% of high-speed users said they were on the Internet at least once per day as compared to 51% for low-speed users. Furthermore, 26% of high-speed users indicated that they used the Internet everyday for business purposes as opposed to 14% for low-speed users. Moreover, 46% of high-speed users consult search engines every day while only 25% of low-speed users. Finally, high-speed users tend to take the Internet with them when they leave their home: 28% of them use the Internet somewhere else than their household as opposed to only 9% of low-speed users.

One other important trend is the steady progression from occasional Internet usage—accessed by dialing a telephone number, then waiting for a series of mysterious tones to "handshake" in order to connect—to the "always on" Internet of second life networks.

Broadband Users are More Aggressive Users of the Internet than Dial-Up Users				
On any given day, the percentage of internet users with each connection who are doing this activity online				
	Broad-band-at-home users	**Dial-up-at-home users**	**All inter-net users**	**Survey date**
Sending/receiving email	59%	41%	45%	May-June-04
Getting news	41	22	27	May-June-04
Checking weather	29	20	20	June-03
Doing job-related research	27	15	19	Feb-04
Looking for political information	21	8	13	May-June-04
Watching video clips or listening to audio clips	21	9	11	Mar-June-03
Banking online	19	6	9	June-03
Instant messaging	17	9	12	May-June-04
Playing games	14	8	9	Mar-May-03
Looking up phone numbers or addresses	12	5	7	Feb-04
Getting maps or driving directions	12	5	7	Feb-04
Creating content and sharing it online	11	3	4	Oct-02
Looking for job information	6	4	4	May-June-04
Looking for a place to live	5	2	3	May-June-04
Participating in auctions	5	2	3	Feb-04
Reading blogs	4	2	3	Feb-04
Buying products	4	3	3	Feb-04
Buying or selling stocks/bonds	2	< 1	1	Feb-04

Comparison of Internet applications as a function of bandwidth

Source: Pew Internet Project surveys 2002-2004

(http://pewresearch.org/pubs/556/why-it-will-be-hard-to-close-the-broadband-divide)

The types of services that have emerged because of high-speed access are emblematic of the growing penetration of the Internet among consumers. In particular, the transition to high-speed access accelerates the consumption of both audio and video content as well as information search and online banking. It is also with the emergence of high-speed access that the first community-based sites begin to appear and allow the on-line creation and sharing of content by its member users. These first forms of social networks are precursors to the more fully developed social networks that will emerge a few years later.

According to a survey[18] from Website Optimization, LLC, in the last five years, the size of the average web page has more than tripled. Between 2003 and 2008, the average web page grew from 94 kilobytes to over 312 kilobytes, some 233%. The use of streaming media on the Web has increased by more than 100% each year. From 2000 to 2005, the total volume of streaming media files stored on the Web grew by more than 600%. While broadband users have experienced somewhat faster response times, dial-up users have been left behind.

High-speed access also allowed for a number of professional activities to migrate online. For example, individuals requiring flexible hours spread throughout the day or week could now work from home or other location. This is possible because they could remotely access information technology previously available only in their company's premises. In his seminal work *The World is Flat*,[19] Thomas L. Friedman used the example of a Mormon family mother in Salt Lake City, Utah, who like her 400 colleagues, is taking reservations for the low-cost airline Jetblue from home. Friedman also speaks of the ability for an enterprise to organize its workflow across time zones: when New-York located employees finish their day, they can transmit their work product to Delhi-based employees who can continue the work as they are just beginning their work day.

High-speed access allows people distant from places of work to find or maintain an economic activity. Most notably this is the case with disabled individuals or senior citizens with limited mobility due to economic or health constraints. In the US, estimates show that the growth in GDP due to the re-introduction into the labor market of senior citizens working from home is above 100 billion dollars[20].

Falling Prices Spurring Adoption

The last important factor for the adoption of new technologies is more microeconomic in nature. The increase in manufacturing productivity caused by the ongoing improvements of electronic components induced a rapid and significant reduction in costs and turned these new technologies into almost everyday household goods.

Never before, in any other sector of the economy, has the famous Schumpeterian process of "creative destruction" been so acutely evident. Under the pressure of competition the development and manufacturing cycles for new components is accelerating and resulting in new and constantly improved products with equal if not smaller size and manufacturing cost. This drives a worldwide and rapid succession of innovation in both the design and features of products and services as well as in their business models. It also continually disrupts the market positions of players including their strategies and processes. Nowhere is this more dramatically evident than in the case of incumbent carriers. Key features of modern industrial history include the *de facto* standardization of components and the simple, rapid, and efficient processes that imposed themselves as a matter of course on the market. These new approaches are based on the logic of mutual interest that emerged from the Open Source software culture (itself a product of the 1960's counter-culture that emerged in California universities).

The growth in the intelligence of electronic components is itself reinforced by the explosive growth in storage capacity it made possible. Just as Moore's Law (discussed earlier) described the exponential growth of the processing power of microprocessors, there is a lesser known law that describes the exponential nature of the evolution of storage on digital media, Kryder's Law, named after the head of Seagate Technology, LLC, who observed that the density of memory on hard disk doubles every year. This is a faster rate than Moore observed for processors. The hard disks first available in 1956 could store 2,000 bits. This has grown 50 million-fold to 100 gigabits (1 gigabit = 1 billion bits) today, an increase well beyond what Kryder's Law predicts[21]. This resulted in an equally impressive price reduction in storage technologies. The retail price in 2003 of one-gigabyte (a billion characters) disk-based electronic memory was \$2.04 and only \$0.77 in 2006[22].

Globally, the reduction in the prices of electronic components does not only affect fixed networks. Mobile networks are also bene-

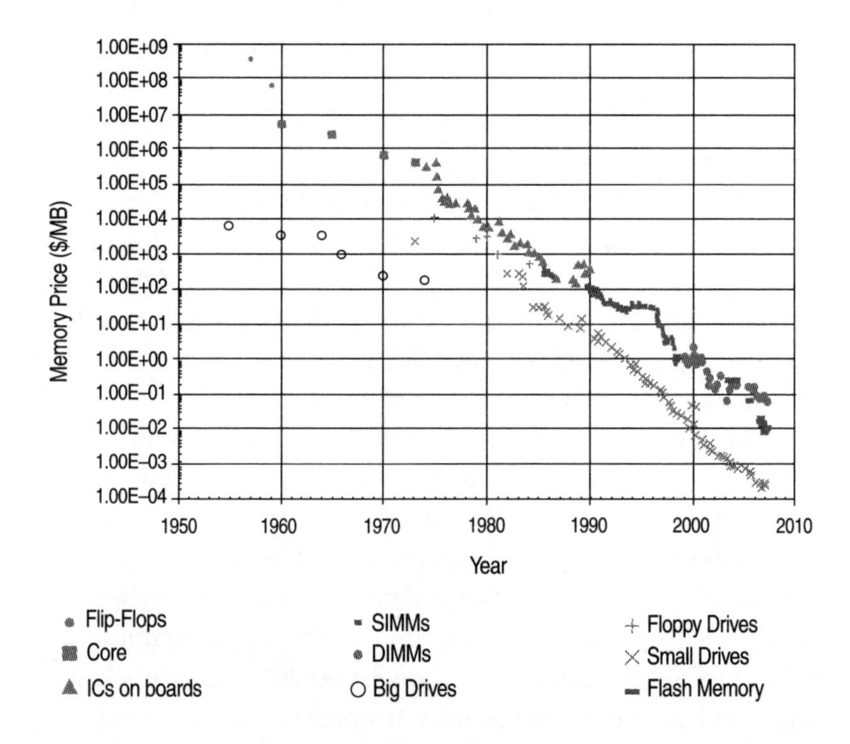

Trends in electronic memory prices

Source: http://www.jcmit.com/mem2007.htm

fiting from this trend due to their own technological evolution and ongoing optimization of the spectrum usage. This is why the penetration rate of mobile telephones has grown spectacularly in just 10 years, resulting in penetration rates in countries like Italy of more that 100% (meaning that individuals have more than one mobile telephone per capita).

As with hardware, software has also increased its capabilities, albeit slower. Wirth's Law describes progress in software thusly: "software is decelerating faster than hardware is accelerating,"[23] which shows that progress in software is certainly nowhere near being exponential. A study by Longstreet Consulting concluded that software productivity has "only" quadrupled since 1970[24]. The reasons for the steep reduction in software prices can be found else where. Key candidates include the outsourcing of code development to low-cost countries such as India or China as well as the increase in freely available open source software downloadable from the Internet.

From a Two-Layer Value Chain...

The gestation phase of the second life of networks also represents a period of profound disruption in the telecommunications value chain. This value chain defines the succession of enterprises that cooperate in the provision of telecommunications products and services to mass market, enterprises, and institutional customers[25].

During the first life of networks, the industrial organization of the telecommunications sector was relatively homogenous and hierarchical. In particular, the strategic nature of telecommunications which relied on a communications infrastructures considered vital to the economic health and security of nations encouraged their governments to maintain a tight control on the sector. The manufacturing of telecommunications hardware was itself organized hierarchically consequently. In the majority of cases the network operators did not manufacture the equipments themselves (essentially switches, telephone sets, cables, transmitters, and amplifiers, based on electromechanical and radio technologies). They were manufactured by essentially captive suppliers with little latitude in deciding the technologies used. Further, network operators conducted the majority of research and development efforts, given their role in specifying their equipment requirements.

The dominant services in this period consisted in renting equipment to enterprises and consumers and providing them with voice telephone service. The notion of content as we know it today did not exist, even if a telephone conversation represented the first instance of user-generated content. Nevertheless, this content left no opportunity for the operator to add value (telephone conversation content was neither stored nor available to be shared with consumers who weren't party to the conversation). In this era, the key performance priorities were quality of service followed closely by the penetration rate of a service among a country's households and enterprises.

Technological development was much slower in telecommunications than in micro-computing. Eventually, equipment manufacturers were able to free themselves from the captive orbit of the network operators[26] and became a separate element in a bi-polar value chain.

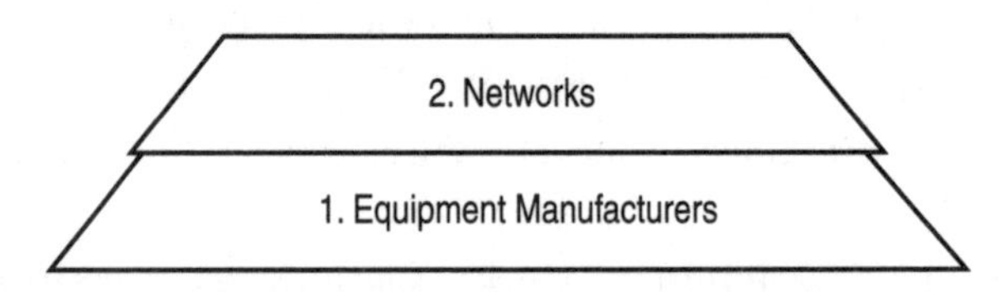

The Two-Layer Model of Telecommunications

There are many instances of this two-layer model. It represented a new worldwide industrial equilibrium between equipment manufacturers and network operators. In the US, there was Western Electric, which in addition to being a traditional electric engineering firm became the equipment arm of AT&T between 1881 and 1995. In France, CIT Alcatel, CGCT and LMT were the equipment providers for the PTT. In Sweden, Ericsson provided equipment for Telia, while in Germany, Siemens, and SEL provided equipment for what is now Deutsche Telekom. GEC, Plessey, and Standard Telephones and Cables (STC) were the key providers for the British Post Office (BPO), the predecessor of BT. NEC provided for NTT's equipment needs in Japan.

An additional characteristic of networks' first life is that the majority of incumbent national network operators had monopoly positions. This policy relied on the notion of "natural monopoly" and held that due to the enormous sums of investment required to build a network, smaller companies could not participate since they had no expectation of success if they were to compete with the incumbent players.

This is why this value chain for telecoms has remained in this bilateral, constrained state for several decades, even as a number of important technology and business model innovations emerged and began to collide at its fringes.

In France during the early 1980's, the Minitel videotex online system, and services, an ancestor of the Internet, represented a signal of the evolution that the telecom industrial organization would face some 15 years later. It created an environment that supported the development of an ecosystem for high value-added services (of a very different nature than the existing simple telephone access service) as well as content. It is interesting to note that even at this early era, the Minitel business model included a free[27] (Minitel) terminal and paid for online textual and basic graphic information services[28]. In this model, the growth in revenues due to usage of the services and content financed not only the free device but also the content and service providers.

In the US, the system of exclusive rental and sale of terminal equipment by network operators came under attack during the 1960's with the Carterfone decision handed down by the Federal Communications Commission (FCC) in 1968. The Carterfone (named after its inventor, Thomas Carter) allowed the connection of a radiotelephone to the public-switched network (in a walkie-talkie mode) using an acoustic coupler[29] (the Carterfone was a predecessor to acoustic coupler modems that would allow the connection of data capable terminals to the telephone network). The Court did not agree with AT&T, the plaintiff, and thus opened the way for connection to its network of a wide range of compatible and non-harmful terminals in addition to those it provided.

Carter had already clashed with AT&T's lawyers a few years earlier, having introduced a device designed to reduce ambient noise during telephone conversations known as the "Hush-a-Phone." He obtained in 1956 the FCC's authorization to sell this device, which, while it attached to telephones, was not in any way electrically connected to it.

... Towards a Multi-Layer Value Chain

It took the technology "Big-Bangs" of the 1970's and 1980's to generate a momentum sufficient to break apart the rigid industrial organization that dominated the telecommunications sector. These events launched the "gestation" period of the second life of networks but also a relatively vast expansion in the telecommunications value chain.

The key technological factors responsible for this evolution in the value chain were the development of micro-electronics and the widespread digitization of information, which greatly reduced the cost of storing, reproducing, processing, and transmitting this information. These advances allowed the creation of digital services and content both of which became available on networks at the time including the Internet. As of 1998, the transport and transmission of digital data already represented more than half of the telecommunications traffic on public networks.

The telecoms value chain quickly expanded to software and information services companies as well as newly created Internet services and content companies including in the latter case companies coming from established entertainment companies. The abundance of content, services, and information created its own complexity. A new type of service intended to deal with this new complexity came on the market: the search engine, epitomized by companies such as Yahoo! and Google and their predecessor, the "Jerry's Guide to the World Wide Web."

As the value chain expanded, the existing network operators also saw their markets open to competition from alternate carriers. Deregulation of telecommunications markets sometimes rushed and uncontrolled, was nevertheless inexorable, and occurred almost simultaneously in most of the developed countries of the world. First initiated in 1984 in the US and the UK it subsequently spread throughout the rest of Europe starting in 1998. By 2001, 79 countries had opened up their telecommunications sector to competition. The "natural monopoly" logic that was the basis of most regulatory regimes led to a number of technical solutions that enabled deregulation. Chief among these solutions was the rental of the subscribers' copper access lines to other operators. There were two modes: partial unbundling, consisting of the sale of wholesale traffic; and full unbundling, consisting of the lease of the complete physical line. In the case of mobile networks, deregulation meant auctioning of the spectrum that each country controlled. The winner (or winners) of these auctions obtained a license to use the spectrum to provide third-generation mobile telephone service also known as UMTS (Universal Mobile Telecommunications Service) to users nationally or, in the case of larger countries like the US, to users in specific regions.

New players emerged to exploit the new opportunities. They either came from other sectors or newly established specifically to do so. This had the effect of breaking the one-to-one relationship between operator and equipment manufacturers that had hitherto existed. Since the existing players were not the only clients of these new players, there was no special influence to exercise on them, as was the case with the equipment manufacturers. Additionally, these new companies evolved in a regulatory context quite different from the strict one applied to network operators. The combined capacity of these new players to innovate and rapidly improve the versions of available services not only created space for new players but also put an end to certain others such as Silicon Graphics, Digital Equipment, Compaq, Burroughs, Univac, NCR, Control Data, and Honeywell. These last five companies were competitors to IBM and called the "BUNCH."

Very early in the second life of networks, the value chain for the telecommunications sector came to consist of four layers: equipment manufacturers, network operators, service providers, and content producers. The various layers of the value chain remain interdependent, albeit not as tightly coupled as were equipment and telecommunications firms earlier. Changes in one of the four layers can dramatically affect the others. For example, Internet service innovation can stimulate the demand for network access; the production of high-definition film content can induce network operators to increase their transmission capacity. On the other hand, eroding margins in telecommunications can lead to changes in equipment sector employment as well as reduce the investment in additional infrastructures required to transmit new innovative services.

As is often the case with technological change, expansion of the telecommunications value chain did not come about easily. It was responsible in the late 1990's for the so-called "Internet bubble" and

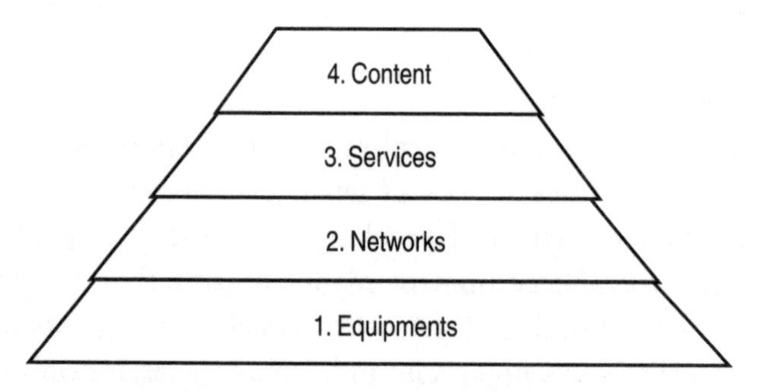

The four-layer model of telecommunications

its attendant "irrational exuberance" (phrase coined by Alan Greenspan, then Chairman of the Federal Reserve).

In a context of constant innovation, it is difficult for players to pursue static strategies. The shocks resulting from the technological Big-Bangs described in Chapter 1 sundered the relationship between past and present. This made it impossible to extrapolate the past in order to predict their economic and financial consequences. Given this structural uncertainty, no investor wanted to miss a potentially highly profitable investment opportunity. This led in turn to a steady climbing positive anticipation of investment returns by the financial markets. This is what explains the high levels of investment observed during the years of the "bubble." The market seemed to have forgotten the simple and prudent fundamental economic rules that could have provided them with a financial "safety line." A related trend that also appeared during this period was an explosion in the number of acquisitions and mergers. This was because companies chose to grow non-organically. They reasoned that they did not have the time to develop the required skills and capital nor to develop themselves the products and services required to compete effectively in the new worldwide markets. Almost daily IPOs of successively greater value accompanied these acquisitions in all of the world's exchanges.

The bursting of the Internet bubble in the spring of 2000 which dragged down the market capitalization of technology and telecommunications firms, followed shortly by the value of more traditional companies, caused considerable damage to financial markets—damage that many at the time considered irreversible. The cascading failures of companies like the telecommunication operator Worldcom that occurred almost at the same time as that of the energy company Enron, as well as that of Global Crossing, only fueled these apocalyptic predictions.

For the US, the precipitous drop in the stock market led to a loss of approximately two trillion dollars of market capitalization. The industrial recession that followed impacted start-ups as well as the established telecommunications giants and resulted in the loss of approximately 500,000 jobs in the US and 1 million jobs in total throughout the world. These remarkable numbers are the result of the over-capacity that developed during the "bubble" and the subsequent cost-cutting measures implemented after it burst (the mergers had led to widespread redundancies in employment that were quickly rectified). Collateral damage also hit the outsourcing sector, as firms in the Internet and telecommunications sector reduced their outsourcing[30].

This period, probably one of the more difficult ones in the history of telecommunications also affected the large equipment manufacturers who had embarked on acquisition sprees as a way to enter quickly growing markets. Some came very close to filing for Chapter 11 bankruptcy protection, as in the case of the fiber optic manufacturer JDS Uniphase. At the same time, traditional equipment manufacturers had to deal with emerging Chinese manufacturers. Only sustained global growth in the mobile sector (driven by terminals) allowed equipment manufacturers, Western and Asian alike, to maintain a semblance of stability during this difficult period.

Yet, only a few years later, the telecommunications sector (and in particular the network operators) have shown remarkable resilience; it was even considered by the financial sector as a safe haven for money during the sub-prime mortgage crisis that arose in mid-2007. The permanent process of innovation and of challenging the *status quo* was primarily responsible for its resilience after the financial turbulence. The revival of the Internet, the growth in mobile services usage as well as the emergence of markets in China, India, and developing countries has allowed the sector to pick itself up again in less than five years.

The Coexistence of "Old" and "New" Players

Beyond the obvious differences between their skill sets, the players in the older layers (equipment and network) and those of the new layers (services and content) distinguish themselves by their capital and organizational structures. While the early equipment manufacturers and network operators had to invest massively in order to provide their products and services, a relatively modest investment was required to participate in the newer service and content layers. This is true even if certain of these players, such as Google, have embarked on ambitious investment paths (see below). Players in the new upper layers of the value chain tend to have very agile organizations and invest almost exclusively in human capital and software. This allows them to innovate and market their services more rapidly than other players in the lower layers of the value chain. The landscape of networks (fixed, mobile, television, and so on) resembles an ocean in which giant cruise ships are attempting to maneuver among numerous and different cabin cruisers and very fast outboard motor boats.

In the same manner, the emergence of China and India is introducing new potential partners and competitors (sometimes both) for Western companies. Major players of the old Eastern Bloc—Russia and the Czech Republic for example—have also recently appeared in the areas of software and customer call centers. Western European and US players are faced not only with production costs that are up to 10 times less, but also with large numbers of highly qualified computer scientists and programmers, particularly in Russia and India.

The same type of coexistence is emerging in the services used, on the one hand, by older telephone and early Internet service users and, on the other, by a new generation of users with radically different approaches and relationships to technology. As William Bailey of the Walt Disney Company remarked in the summer of 2006: "Previously, there was a recurring debate around whether it was content that was king or rather it was distribution that was queen; now it is the consumer that is king."[31]

It is interesting to note that search engines like Google's have not completely replaced telephone directories, which are still published and distributed. Nevertheless, early in the 2000's, yellow page directory companies began to develop sophisticated Internet sites and tools as a way to capture advertising revenues spent by clients on other sites or on search engines that represented a sometime more efficient alternative. This is an important example of how first-life networks players are orchestrating innovative strategies to improve their services and accelerate their entry into their second life.

Similarly, in the area of technologies, the "open" philosophy of the Internet and the "closed" environment of the mobile telephone are coexisting. On one hand, the Internet flourished in an essentially deregulated context. Since it is made up of a network of networks, no single company totally controls the Internet. As put aptly by Ori Brafman and Rod A. Beckstrom in their book *The Starfish and the Spider*, "[there is no] President of the Internet."[32] This is why telecommunications operators have invested in new areas like Internet portals, email, and e-commerce even if it required development of different skills. In the Internet world, controlling terminals has little importance since in any case operators have no monopoly on them.

In the mobile world, however, the economic rules of the game are different and, given the extreme complexity faced by users to procure directly handsets on their own, the whole value chain is vertical, that is, managed by the network operators. They purchase terminals and resell them to users at attractive prices subsidized by revenues from prospective communications service use.

Radically different network pricing models coexist. For example, the traditional pricing scheme for wire-line and wireless telephone service, which includes a fixed subscription component plus a usage-based one (from the first minute with wire-line and after plan minutes are exhausted in the case of wireless), coexists with an "all-you-can-use" flat-rate model used for the Internet, Voice-over-IP and TV-over-IP. In the end, the coexistence of these different business models is what marks the end of this "gestation" period. Today, both operators that bill clients directly and Internet services companies that capture advertising revenues indirectly are flourishing. In the latter case, there are no contractual commercial relationships with the end-users consuming their services who are more like their "audience" than their "clients." Indeed, many expect that this type of model will dominate the industry in its second life.

In summary, the first life of networks, however archaic it may seem in some of its aspects, did not completely fade away; in practice, all of its components remain today in one form or another.

*

In 2004, while Google made a spectacular splash with its IPO and became an icon for new Internet players, the new technologies created by the "big bangs" of the 70's and 80's had become widely distributed and understood. All the conditions necessary for an explosive growth in the use of available services as well as the total immersion of users in networks were finally in place.

The constellation of physical and social networks that had emerged was finally poised to start its second life.

CHAPTER 3

We Are the Networks

The year 2004 marks the symbolic birth of telecommunications networks' second life. In what follows the relationship of users to the new information and communications technologies that began emerging at the end of 1990 will be characterized.

These new technologies made available to the market (what the economists call "supply") performed reliably and for the most part were sold at affordable prices. Naturally, corresponding adoption by consumers (or what the economists call "demand") also developed. This originated from two sources. First, from the increasing adoption by analog fixed-line telephone users of Internet and mobile telephone services. Second, from users' growing mastery of the technical aspects of these services and their associated devices. The telecommunications value chain followed closely these shifts (and in some cases even anticipated them relatively successfully), negociating effectively the resulting unprecedented changes in its industrial organization.

However, if we look a little closer, the essential defining characteristics of the second life of networks are much more profound. It consists of immersive telecommunications, omnipresent and ubiquitous networks, and finally, the merging of individuals with these networks. Networks in their second life can be considered as the counterpart to the Web 2.0 concept[1], which came on the Internet scene in late 2004. If Web 1.0 facilitated access to information, Web 2.0 made the availability of user-friendly, interactive Internet services ubiquitous.

Life before the Internet?

Source: Pirillo and Fitz. (www.blaugh.com)

Within ten years, users throughout most of the industrialized countries have come to consider information technologies as fundamental (we cannot do without them anymore) rather than simply as utilitarian (useful for searching for information or for contacting someone else). How many times have people asked themselves, "How did we do without the mobile phone or the Internet"? In a word, we have entered the era of "always on."

So, the start of networks' second life is a milestone as important as the birth of the Internet and the universal availability of mobile telephones ten years earlier. The end of the 1990's saw the appearance of new telecommunications technologies and the beginning of their adoption. The end of this decade also saw the beginning of an era of permanent connection of individuals and enterprises to networks. This was made possible by these new technologies and the resulting enthusiasm of investors as well as consumers for the new services they enabled. This growing enthusiasm also led to visionary and even sometimes unrealistic expectations.

It remains a mystery how the corpus of telecommunications literature—be it technical, economic or business—ever prolific regarding the first Internet bubble's causes, implications, excesses, and disappointments, has demurred on the reality, the stakes and the tremendous opportunities of the "post-bubble" world. It is as if an "amnesia of the future" caused by the traumas of the post-Internet bubble world has installed itself in the collective memory.

Unexpected Uses of Common Technologies

The immersion of users in the new world of information technologies is due to their successful appropriation and mastery during the development phase of second life networks that occurred between 1995 and 2004.

This is quite concretely demonstrated in the case of mobile phones where its use has become virtually instinctive, with little if any reference to an instruction manual, imperceptibly drawing the user's attention, and becoming automatically, even seamlessly integrated into the user's daily routine. The mobile telephone in 2000 was a little like the wristwatch in 1930: adopted by almost everyone, a new accessory worn permanently and naturally consulted throughout the day without interrupting the flow of one's routine. This analogy is complete when we observe that the mobile phone today is used by many also as a watch. Similarly, connecting to the Internet in 1995 disrupted a household's routine since it would disable the telephone except in the rare case where there were two telephone lines. Today, users do not "connect" to the Internet. In most cases, they are automatically and permanently connected. Being able to "surf" the Internet "seamlessly" has become indispensable to these users.

Beyond these direct implications of the developmental period of networks' second life, certain structural aspects explain the advent of "always on." The rich functionalities and features of the new technologies lead not only to their frequent use but also explains the growing immersion of users in networks. The fact of the matter is that networks have become indispensable and omnipresent because they enable with little effort, rapid, and simple responses to the daily needs of private and professional life.

Internet and mobile telephone services have progressed from being luxury goods in the 1990's to consumer goods in 2000. Today, in the era of second-life networks, they have become almost essential goods.

Fixed-line voice communications had a major impact on the world economy as well as on social and cultural life everywhere. Users became accustomed to the telephone as an essential service without any new or radical innovations changing the basic way it was used. The traditional telephone conversation of the third millennium remains almost identical to the one at the end of the 19th century. This is in spite of the fact that its quality and reliability have largely improved, its prices fallen considerably, and its pricing less based on call length and distance. Individuals use Internet and mobile services in many different ways, however. For example, as the Internet grew, driven by the complementary growth in personal computers, it created new and unexpected uses. The PC was originally conceived as an "offline" (unconnected to a network) experience, relying on multimedia CD-ROM's for content. Microsoft was at the time the major proponent of multimedia CD-ROMs. In a cautionary message to his employees on December 7, 1995, Bill Gates announced: "The tidal wave of the Internet [is here]: the sleeping giant has awakened!" paraphrasing Admiral Yamamoto's declaration after the Japanese attack on Pearl Harbor: "I am afraid we have awakened a sleeping giant."[2] At almost the same time, James L. Barksdale, the CEO of Netscape, which was Microsoft's main competitor in Internet technologies remarked, candidly and confidently, "God is on our side..."[3]

The rapid spread of the mobile telephone, for its part, freed users from the need of being near a specific location in order to place a call. The mobile telephone has transformed a state of location-based dependence into one that evokes a sense of belonging to one's personal communication network and in so doing set the stage for an expansion of related applications.

One aspect of this proliferation of new uses is that a significant part of them were not anticipated by experts. The most representative example is SMS (sending text messages between mobile handsets), whose tremendous growth was unanticipated and unexpected. However, on New Years Eve 2007 in the Philippines—"the text messaging capital of the world—a remarkable 1.39 billion text messages were sent from a subscriber base of just 50 million". There were an estimated 43 billion text messages sent worldwide the same evening[4].

The Gartner Group estimated that there were 1.5 trillion mobile messages sent in 2007 in Asia/Pacific and Japan, 189 billion in North America and 202 billion in Western Europe. Similarly, no one completely anticipated the mass-market success of email.

In practice, consumers themselves discover or create these new, unforeseen applications by diverting the service or hardware from its original purpose. This is referred to sometimes as "hacking," a word originally having illegal connotations which is today also used to mostly denote "unsolicited" creativity. This phenomenon of unforeseen emerging uses is not without historical antecedents in telecommunications. For example, during the high school qualification exams of 1967 in France, students used their telephone in an interestingly unconventional manner. They knew that on dialing a sequence of numbers or letters, the call would be routed to a message announcing that no such number existed. During the silence between the repeating message sequences, the caller could speak to other people who had performed a similar dialing procedure. There was a rumor that students could get a preview of the exam questions with this use of technologies. In the US, a similar "hack" of telephone service called the "party line" became popular. It however had nothing to do with the rural party line services that were legitimate offerings of telecommunications companies.

But perhaps what is most indicative of the shift marking networks' entry into their second life is that services already available for several decades are finally beginning to take off. For example, it was widely believed in the 1970's in the US that users would rapidly adopt the "Picture-Phone."[5] It was not until the availability of smaller, low-cost "webcams" that could be connected to or integrated with a personal computer that the service gained in popularity. The decent quality of webcam image definition allowed by the growing transmission capacity of networks is one example of a simple and "seamless" use of telecommunications.

The Emergence of Social Networks

The immersion of users in networks is a consequence of their adoption of the new technologies. The number of ways networks are used is broadening and signals the advent of a new mode of communication. It is characterized by a progressive shift from the traditional "one-to-one" telephone call to a "one-to-many" and "many-to-many" mode of communicating. A telecommunications engineer would describe this as moving from a "point-to-point" to a "point-to-multipoint" and "multipoint-to-multipoint" transmission architecture.

At each instant, almost effortlessly, users can thus communicate with their friends, their families, their professional colleagues, and social organizations they belong to, in a word, with their networks. Networks' second life is marked by the fusion of physical telecommunications networks with those that will be described henceforth as "social," "human" or "community" networks. Omnipresent and ubiquitous physical networks are being complemented by a similar expansion of omnipresent and ubiquitous social networks. These are allowing users to continuously inform members of their networks of events in their private and professional lives, their tastes, preferences, and so on. It is likely that in the near future physical communication networks will emulate the way today's social networks organize themselves and cooperate to generate real time context sensitive "decisions."

These can be considered as a new mode of communication because social networks are now supplanting traditional modes of communications, and in particular, physical ones. Among young Americans, face-to-face meetings represent only 35% of their communications while communicating through online social networks accounts for 47%[6].

The sheer amount of available online information and content is outdistancing individuals' capacity to be even aware of its existence. Users have transitioned from "a situation where attention was abundant and content rare to one where attention is rare and content abundant."[7] Thanks to a variety of new technologies, the "physics of the media world" are changing. With so many competing options, the individual's attention is now the most valuable commodity.

The popularity of social networks lies in the simplicity of their use and their low cost. They allow users to "multi-task" and simultaneously manage all of one's relationships, much like Napoleon, known to have dictated five letters at a time. It is now possible to inform geographically distant family of a child's first tooth, while simultaneously checking the availability of fellow team members for a volleyball game, and sending an important article to work colleagues. There are also advantages that go beyond these basic ones: online social networks allow access to groups not otherwise reachable. Cooking is a good example of this. While a hobby for many people, it is nevertheless a time consuming one to truly master. As a result, there are many online sites that have been created and that provide recipes, recipe variants, cooking tips and meal wine choices. Most of them are experiencing exponential growth in the number of visits they receive.

This new mode of communications is sometimes referred to as "semi-synchronous" because it is not completely synchronous like in the case of a telephone or a face-to-face conversation, nor asynchronous as in the case of email (since a response need not always

The Diversity of Social Networks

Drawing by Marc Fersten, 2008.

be deferred). As pointed out by Linda Stone, writer and consultant, formerly of Apple and Microsoft, SMS is the best example of this new mode of communications which is "usually used in a 'semi-sync' mode."[8]

The new methods of social exchange made possible by the invention of the telephone were an important factors in driving its adoption. The system required passing through human operators in order to be connected to the correspondant as well as to get directory information and even merchant suggestions if needed.

A parallel development in the "socialization" of telecommunications networks was taken with "CB" or Citizen Band Radio. Starting in the 1950's in the US, the CB allowed individuals to communicate using low power, limited range emitting-receiving radios. Since, CB has remained primarily an application for roaming individuals such as truckers, taxi drivers, and couriers. The conversations are on open frequencies so that everyone in a band's operating range can hear the exchange and initiate discussions with everyone else, whether they know each other or not.

With the development of text-based electronic communications and its subsequent availability to the mass market, community-based usage rapidly grew. The first significant example of this new trend was the Minitel in France in the 80's (which enabled, in addition to other services, "social messaging" forums as they were originally known). It joined a growing list of similar companies, including Prestel in the UK, Bildschirmtext in Germany, Telidon in Canada, Captain in Japan, as well as Viewtron and Prodigy in the US[9,10].

At about the same time, with the growing power of communicating personal computers, a new generation of micro-servers were developed which allowed BBS (bulletin board services) to flourish.

A Visionary Cover of Computer Magazine (1979)

Source: cover of "COMPUTER Magazine," September 1979. (http://www.computerhistory.org/internet_history/internet_history_70s.shtml)

These were managed primarily by avid BBS amateur hobbyists. These new servers allowed the dissemination of information over the network and supported real-time as well as store-and-forward messaging services and various other applications such as a free software exchange. The term "Bulletin Boards" was a good way to describe these services and remains an apt metaphor even today for what social networks have in part become: a place, a virtual board of sorts, where its members can post their messages and interact with each other, individually or in groups.

With the development of the Internet, electronic messaging, forums, and blogs quickly became leading applications and gave a social dimension to electronic communications. But the true pioneers of virtual communities as we know them today were The WELL (launched in 1985 as BBS, then as an Internet site), followed by Tripod (launched in 1992), Geocities (in 1994), Classmates.com (in 1995), and SixDegrees.com (in 1997).

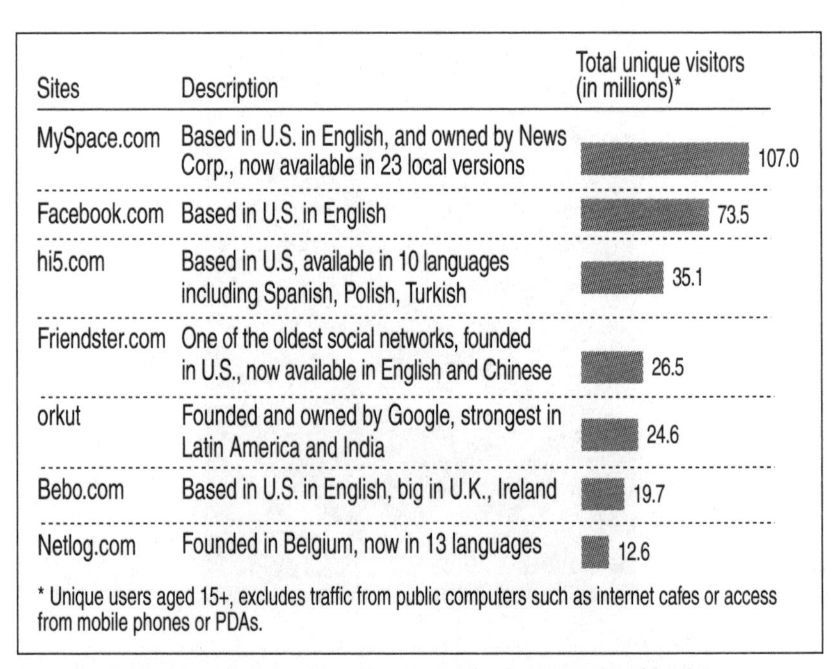

Sites	Description	Total unique visitors (in millions)*
MySpace.com	Based in U.S. in English, and owned by News Corp., now available in 23 local versions	107.0
Facebook.com	Based in U.S. in English	73.5
hi5.com	Based in U.S, available in 10 languages including Spanish, Polish, Turkish	35.1
Friendster.com	One of the oldest social networks, founded in U.S., now available in English and Chinese	26.5
orkut	Founded and owned by Google, strongest in Latin America and India	24.6
Bebo.com	Based in U.S. in English, big in U.K., Ireland	19.7
Netlog.com	Founded in Belgium, now in 13 languages	12.6

* Unique users aged 15+, excludes traffic from public computers such as internet cafes or access from mobile phones or PDAs.

Top Social Networking Sites in September 2007 (worldwide)

Source: *The Wall Street Journal*, September 2007.

New generations of social network services have since appeared with MySpace, Facebook, Friendster, Orkut, Bebo, Hi5, Xanga, as well as others, and have experienced significant growth since their inception in 2002. These social networks all have, more or less, the same features and *modus operandi*. New users are invited by an existing member through a host site that sends an invitation email. Frequently, the email addresses are extracted from members' address book at the time they register and the invitation emails sent automatically. Few users are aware that in fact they are allowing the social network the latitude to invite, without their explicit or conscious approval, not only their friends but also their business contacts. Moreover, the fortuitous rediscovering of "friends" (the term used by Facebook to describe members of a social network) is not always something wished for by the user. These occasions are sometimes marred by untoward motivations such as for example the desire to obtain information for business advantage by the person initiating the contact.

As new members join the social networks, the same process is repeated resulting in a "snowball" effect. This explains the viral growth in the number of users often associated with such sites. Moreover, this is achieved with little or no cost of customer acquisition. This viral effect is tied to the idea that "the world is small." The psychologist Stanley Milgram demonstrated in 1967 that the number of links required to connect socially any two individuals in the world is generally quite small. According to his studies, the average number of these intermediate social links is five, which is the basis of the social networking concept of "six degrees of separation."[11]

Beyond the simple connection of people, online social networks also enable users to post their "profiles" in which they can describe their personalities, tastes, and preference. This information allows the identification of others individuals with common affinities. This profile is often elicited and developed across a range of entertaining and engaging applications. One such example is a film quiz application that reveals and codifies users' film tastes and preferences. The resulting indexing of users' tastes allows easy identification of like-minded users. Naturally, the public posting by users of their personality, tastes, even sometimes their political affiliations, religious convictions and resumes, attracts the intense interest of recruiters and advertisers. This, however, poses some major questions about privacy.

Once members of a social network are "profiled," users are connected permanently and fluidly with members of groups that are of interest to them: family, friends, online "friends," professional contacts, sports team, etc. Each member can follow even the most minor changes in the experience of other members of the group without the underlying physical network imposing itself by requiring a special effort of any kind, such as establishment of a call for example. In fact, users do not even have to go and get the information; it comes to them as soon as it is available. This profiling of individuals is akin to the indexing of information. Indexing allows information to be characterized and matched; profiling allows a similar operation on individuals. Moreover, networks have progressively enabled users to share content they have produced or created. This is called User Generated Content ("UGC") and can range from photos, videos, music, and text. More recently, user generated services and applications have begun to appear.

Several hundreds of millions of individuals participate in social networks today. This has resulted in their two and in some cases, three-digit growth rates. Social networks for children have even emerged[12]. Certain of these social networks are regional and some even dominated by users from one country. Brazilians, for example, were the primary users of Google's Orkut in 2007. Other social networks are specialized in travel, pets, art, football, golf, cars. In any case, each of these social networks develops its own differentiating focus reflected in the type of content shared or the information made public about other users.

MySpace, for example, contains primarily pictures and music and represents an important vehicle for discovering the music of independent artists. Certain famous groups like U2 and Coldplay have created a page on MySpace, realizing that its users, most relatively young, are an excellent target audience. Facebook, originally limited to students and alumni of universities is now open to all, began by emphasizing text-based communications. Over time, it has also become an important platform for storing and disseminating pictures videos, as well as applications developed by its members (who include smaller software development companies); this has contributed to its increasing popularity and has attracted large numbers of users. Facebook claims it is the top photo and video sharing application on the Web (according to comScore, its photo application draws more than twice as much traffic as the next three sites combined with more than 14 million photos uploaded daily[13]).

In early 2008, Facebook is arguably the most fashionable social site, growing at a rate that could allow it to catch-up to MySpace, the leading social networking service. This dazzling success (Facebook was created in 2004), is attributed by its founder and CEO Mark Zuckerberg to exploiting to the fullest degree possible what he calls the "social graph" of its members. This social graph allows the tracing of the interrelated branches of its members' social relationships as a function of their interests and preferences. In particular, it characterizes the strength of the relationship of each link as a function of its total number of linked social connections[14].

For its part, the social network Second Life is probably the most representative of the second life of networks. On the majority of social networking sites, members appear, if not always under their "real" identity, at least as Internet users. More importantly, there is a presumption they have an actual existence somewhere on the planet. Members of Second Life start by creating a customized avatar, more or less human looking, and make this avatar evolve as they please in a completely parallel world. Once created, the avatar need not represent its creator: their respective evolutions can be separate. An avatar, for example, can start a business and earn revenues (Second Life has its own monetary system) while its author may not earn any money in the real world. In Second Life, the underlying physical network completely disappears from the user's experience (even if it remains essential to allowing the experience) in favor of a visible virtual social networking environment[15].

The social network LinkedIn facilitates the identification by its users of potential business partners, intermediaries who can provide references, recruiters, and "Mavens,"[16] a category of key influencers identified by Malcolm Gladwell[17]. In practice, each member of the social network makes public those parts of their professional resumes they deem desirable.

Social networks have also begun to appear within enterprises. Since 2007, companies like IBM, Microsoft, and Cisco have provided hardware and service solutions to companies based on the concept of Enterprise Relationship Management.

The User as a Node at the Crossroads of Multiple Networks

The user has gone from playing a passive role as a "cold" endpoint of a bilateral or point-to-point physical network to playing an active role, a generator of "entropy,"[18] as a "node" in a mesh of numerous social and physical networks: henceforth, *we are the networks*. It is interesting to see how far the physical communications networks have been sublimated inside the social networks as Facebook and Myspace. At the TieCon conference in San José, California, in May 2008, Chamath Palihapitiya, Vice President of marketing and operations at Facebook, said "... We're the cable company creating the pipes, and what they carry is social information and engagement information about people," he said. "That plumbing should exist around the Internet, and then what happens is that people can create truly social experiences on the Web. As long as we can build the plumbing for this, we view that as a success". This shows if need be, how intertwined physical, virtual and social networks have become.

By linking themselves to their social networks, users are doing nothing less than unconsciously applying the principle of thermodynamics: nature tends to increased entropy, favoring disordered states over ordered ones, complexity over simplicity. In the end, the

From Edge to Core, From Passive to Active: Users at the Center of Multiple Networks

Drawing by Marc Fersten, 2008

success of social networks is due to the fact they follow the same path as nature, the same path as history.

On one hand, becoming a "node" in such a mesh of networks gives the user a measure of satisfaction and well-being, even recognition as well as ultimately "utility" in the sense used by economists. Linda Stone[19] explains: "To pay continuous partial attention is to pay partial attention—*continuously*. It is motivated by a desire to be a live node on the network. Another way of saying this is that we want to connect and be connected. We want to effectively scan for opportunity and optimize for the best opportunities, activities, and contacts, in any given moment. To be busy, to be connected, is to be alive, to be recognized, and to matter."

On the other hand, the complement of this increased utility accruing to users is that more than ever, the value for telecommunications operators of their networks is in their nodes.

Mathematically speaking, the emergence of social networks revises the traditional method of modeling the value of networks as a function of the number of its users. The canonical example of this approach is Metcalfe Law, which calculates the value of a network as proportional to the square of its users, that is to say, as a function of the total number of users taken pair-wise. This law works very well with telephone networks, where it is intuitively obvious that the number of bi-lateral conversations it allows reflects the correct value of the network. *A fortiori*, Sarnoff's Law, which calculates a network's value as directly proportional to the number of users, is less applicable (being more relevant to electrical, gas, water and television physical networks or grids where each user is in a one-way relationship with a collectively consumed resource).

In the second life of networks, users do not just simply have the choice of entering into bilateral contact with other users. They can also choose to initiate communications with a number of other users in whatever combination they wish and as a function of their affinities. The value of a network with these possibilities is based on the combination of sub-groups that can be constructed among all the users. This can be represented by N, which results in the expression 2^N: known as Reed's Law developed in 2001[20].

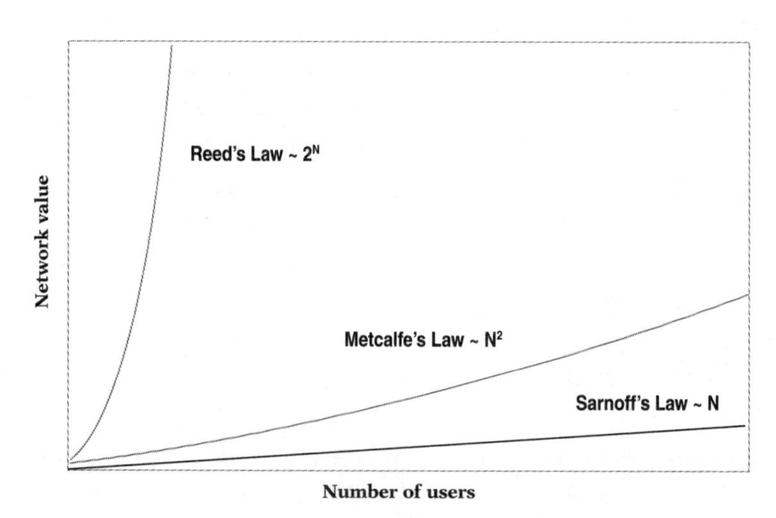

A Comparison of
the Laws of Sarnoff, Metcalfe, and Reed

The traditional limit of Metcalfe Law, which physically constrains the number of people a user is in contact with, has been made largely irrelevant in its current form, at least when it comes to today's networks. Social networks, through their numbers and popularity, have significantly increased the number of people an individual can be in contact with.

Nevertheless, it is obvious that Reed's Law also has its limits: it is humanly impossible to be in permanent contact with hundreds of sub-groups, all with different interests. In reality, the exponential curve describing Reed's Law will flatten out at a certain critical value equal to the number of people a given individual could possibly interact with realistically.

The lower boundary of this number is most likely to be found in the work of British anthropologist Robin Dunbar. His research suggests the concept of Dunbar's Number which describes the cognitive limits of the brain's neo-cortex in matters of social relationships. While there is no precise number which can be calculated, as the case with the various other laws mentioned, 150 is the estimated number of people with which an individual can have a stable relationship[21]. This number is at best a lower limit value since it simply considers the native capacity of a human's, or for that matter, a higher-order animal's, neo-cortex. In particular, it does not consider the new technologies available to humans in the form of social networks. These can significantly enhance an individual's capacity to manage and expand the number of relationships they pursue, if they so desire. Nevertheless, this number is not without bound.

New Generations for New Technologies

The growth in social networks—which until now has been characterized as the result of the ease of use of their underlying technologies—has also been accelerated by the new generations of "ultra-technophiles" who are their principal users.

These new generations have a number of designations. The most popular include the "Net Generation" and "Gen Y." They are also known as "Millennials," the "Echo Boomers," and the "Connected." They are also sometimes called the "Me Generation." "These generations have been educated in a system that largely emphasizes self esteem rather than material accomplishment"[22].

Inter-generational incomprehension

Source: Special to the Connection. Adam Is © By Brian Basset, Universal Press Syndicate

Even if there is no strict definition for Generation Y, it consists generally of men and women between the ages of 18 and 30, born during the technological transition between the mid-70's and 90's. They grew up immersed from a very early age in the new information and communication technologies and unlike generations born earlier, did not have to go through a phase of progressive adoption and mastery of these technologies. This unconscious, tacit adoption of technologies often generates inter-generational tensions and incomprehension between the "offspring" of networks' first and second lives.

Generation Y consists of about 70 million of the some 300 million US residents. They have significant purchasing power and influence relative to other generations, primarily in terms of tastes, behavior, and consumption. This economic power comes primarily from their numbers: Generation Y is the offspring of the baby-boom generation. They represent, like their parent's generation before them, a noticeable spike in worldwide developing countries' birth statistics.

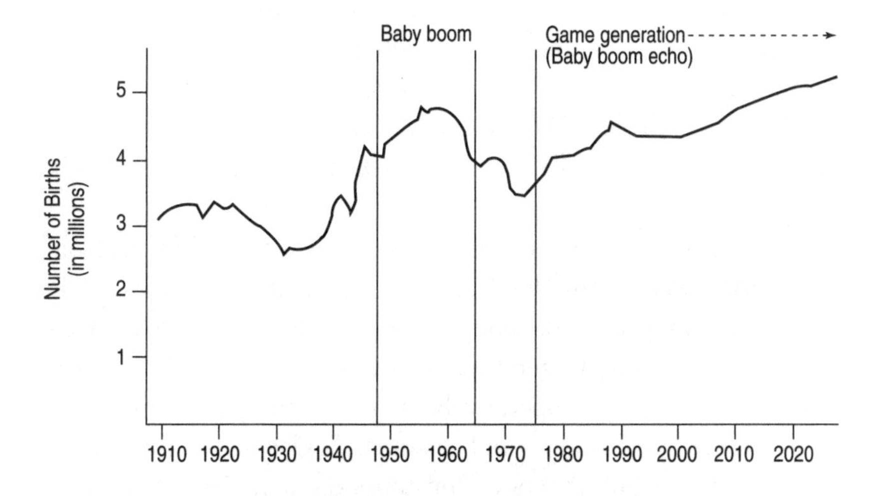

Demographic importance of Generation Y

Source: Beck J., Wade M., Got Game: How the Gamer Generation Is Reshaping Business Forever, Cambridge (USA), Harvard Business School Press, 2004, p.17.

Beyond their "ultra-technophile" characteristics, what other factors describe more precisely Generation Y's new technology consumption patterns? Five key factors emerge from quantitative studies of its members.

—First, Generation Y members maintain a *physical proximity* to their telecommunications networks (naturally though their devices). This ensures that they are "always-on," "multi-tasking," and "multi-mode" (fixed, mobile, internet and TV). As an illustration, a study done by China Mobile found that 91% of young Chinese users of mobile services keep their telephones within a radius of one meter, 24 hours a day and 7 days a week. For these new generations, mobile telephones represent a very personal and private accessory, earning them the label "the mobile generation."[23]

—Second, Generation Y members' are unconsciously immersed in physical networks and are regular users of Internet *social networks* like MySpace, Facebook, Twitter, or Second Life, as well as community-based information sites like del.icio.us or Digg. These permanent online contacts with their social networks have become for Generation Y a more important way to pursue personal interactions than physical or traditional telephone-based contacts. Deloitte Consulting found that in 2007, Generation Y members have on average 37 contacts in their IM (Instant Messaging) "buddy-list" as compared to only an average of 17 contacts for the population as a whole[24].

—Third, Generation Y's consumption style is very different from that of prior generations' due largely to having grown-up in a world of *impatience* and "instantaneous response."[25] It is therefore no accident that Generation Y members tend to be adept at video games, which they often play cooperatively or competitively, over the Internet. The underlying motives and gestures of video games themselves are characteristic of the new consumption patterns. In a video game, for example, the fact that players have multiple lives (that "death" is temporary) leads them not to think twice about using a tactic with uncertain outcome since the cost of failure is low. Failure indeed can become a viable learning strategy. This is why, when faced with a new Internet service or software, Generation Y users will try out in an inductive and sometimes impatient fashion, all of the features or content available in the hope of finding the things that work best for them. This is also true in the way they discover new hardware technology: after all, if this peripatetic and inductive approach to discovery "locks up" the PC or gaming platform, it can always be re-booted, with little adverse consequence. Previous generations, however, are used to a more deductive and staid approach to using new technologies. For example, an error in dialing a string of 10 numbers on a dial-telephone required the user to begin again, to pay for the communication if it connected—and most of all to have patience!

—Fourth, Generation Y *participates* more actively in the life of social networks than the previous generations by contributing to their content by its own creations, the UGC. In 2006, 33% of Americans between the ages of 18 and 24 had already used their mobile telephones to take and upload a picture to social networks[26]. Of course, User Generated Content (UGC) is not limited to pictures only and can include videos, music, and text as well as SMS's, in some cases.

—Finally, members of Generation Y are *entering the work force* today with an excellent, almost visceral understanding of communications and collaboration tools (IM, software downloads, blogs, wikis, Internet video, and the like). In fact, they are not shy about bringing into their workplace technologies that are more advanced than those used by their companies. It is this factor that most distinguishes Generation Y from others in the enterprise.

The key challenge for the telecommunications sector—which, more than the other sectors of the value chain, needs to be ahead of its time—is not so much to adapt to the Net Generation. Rather, it is to prepare for the one just behind it, whose members' are even more immersed in networks then preceding ones'. For members of this emerging generation: the First World War began 80 years before their births; they never knew the USSR; there has always been only one Germany; they missed the May 1968 Paris student protests riots; they never experienced man's first steps on the Moon, the assassination of JFK, nor the accidental death of Elvis Presley. These events all occurred a quarter of a century, more or less, before they were born. They have never seen a camera with rolls of film, nor can they imagine a world without mobile phones, DVD's or the Internet, and their address books include friends throughout the world.

According to a study by the Pew Internet & American Life Project, in November 2006, in the US 93% of adolescents (ages 12 to 17) were connected to the Internet, 61% using it everyday and 34%

several times per day. Two-thirds of them created content on the Internet and 55% used online social networks; this percentage jumps to 70% in the case of older girls of the cohort. The major activity of adolescents in these social networks is communicating with other members[27]. For example, 84% of them post messages on the personal pages of their friends and 82% send them messages *via* social networks[28].

The percentage of teen Social Network Service users who...	
Post messages to a friend's page or wall	84%
Send private messages to a friend within the social networking system	82%
Post comments to a friend's blog	76%
Send a bulletin or group message to all of your friends	61%
Wink, poke, give "e-props" or kudos to your friends	33%

Margin of error is ±5% for teens who use social networking sites. Teen social networking site users n=493. Table first printed in Social Networking and Teens data memo, available at http://www.pewinternet.org/pdfs/PIP_SNS_Data_Memo_Jan_2007.pdf

Adolescents and social networks

Source: Pew Internet & *American Life Project Survey of Parents and Teens*, October-November 2006.

Beyond the Search Engine: Recommendation, Redundancy and Reputation

"We are the networks": this is the quintessential motto of the second life of networks, especially for Generation Y and the generations that will follow. This new status of users as a network node is not limited to only consuming services and producing personal content, however.

It also implies a very particular and novel role. Members of networks can also now act as "guides" for their peers to the abundant content made available by these networks in expressing personal preferences, advice, evaluations, and recommendations. In a sense, users are, in networks' second life what human operators were, in their first life[29]. They guide their fellow network "travelers" to areas of information and content they have already first explored. This is a more subtle and complex role than that of search engines which, as intelligent as they may be, remain automatic, artificial and are in particular, unable to adapt to unanticipated queries or understand their "sense" or semantic. To the axiom "we are the networks," we can add a corollary: "it isn't what you know but who you know." Taken together, these aphorisms describe the roadmap to the second life of networks.

It is thus that the most advanced Internet users—those most prolific in expressing their opinions online about the latest film they saw, the news or even food recipes—have established a sort of "fifth estate" of second life networks. The zeal with which the candidates for the US 2008 presidential elections tempt to seduce these Internet users is completely symptomatic of this emerging class of influencers.

However, even if users belong to the same social network, how can the opinion of another unknown user be trusted? After all, they may live at the other end of the world and have tastes and preferences that are very different. This fundamental question in the second life of networks calls above all for a change to the way one searches information. Henceforth, it is the abundance of information itself, the ability to cross-check sources, and its redundancy that will ensure its quality. If someone is interested in a new film, they can read a dozen opinions penned by Internet users in their social network or not, and factor out the most common denominator of the opinions instead of depending on only one opinion that may be subjective and incomplete.

In his book *The Wisdom of Crowds*, James Surowiecki shows that the aggregation of information in a group leads to decisions that are often better than those taken by a single member of the group. One of the applications of the "wisdom of crowds" which was called "common sense" in simpler times, can be found in prediction markets which resemble a sort of an "opinion exchange" for current affairs For example the question "who will win the US 2008 election?" represents an instance of the buy or sell order[30].

This mode of interaction with networks is reminiscent of the technical mode of data transmission over the Internet described earlier: the IP or Internet Protocol based on the principle of "best effort delivery."

This technical choice simplified and optimized the equipment responsible for "routing" the information being transported on the network. This routing equipment (routers) would make their "best effort" to send the "packets" of information (also known as "datagrams") assigned to them on to their next destination. This equipment would do its best to ensure that the data arrived integrally and within a reasonable timeframe, but without any guarantee. Each router sends to the next router "it can see" the packet it received without being certain whether it will be correctly received or sent on. This lack of certainty is precisely the reason why it is impossible to commit to "quality of service" guarantees on the Internet, that is, to ensure close to 100% reception of the sent information. Furthermore, the routers treat each packet of information sent independently of the other packets of the same message, be it text, images, sound, video, or data. Each message packet can reach its destination across different paths in the network, which requires re-assembly of the message at its final destination node.

This type of transmission is radically different than the one used in first life telephone networks based on the "circuit switched" paradigm. In this mode, the communication can only proceed if a circuit path between the caller and called party is pre-established. Quality of service was therefore guaranteed, which is the hallmark of the circuit switched paradigm.

The underlying protocol that allows the "best effort" principle and makes it reliable by rendering any associated packet loss imperceptible to the user is the routers' capacity to resend lost or damaged packets as well as to change the paths used as a function of the network's state, that is, its traffic pattern.

By analogy, on the Internet, the abundance of user generated content sources and the range of ways to access them—blogs, wikis, personal pages, or web sites—play together the same role as the retransmission of IP packets. It ensures that beyond a certain level of irreducible uncertainty; a user can be convinced of the veracity or viability of information obtained by verifying and comparing it as many times as necessary with duplicate sources of the same information.

In practice, users either can directly or with the aid of content aggregation tools found on the Internet, evaluate the probability that the information they obtained is correct. One critical measure of the efficiency of verification by progressive convergence (or trial and error) is the number of times an alternate site[31] needs to be consulted before the user is convinced of the relevance and reliability of the information initially found. Equilibrium is reached when consulting an additional site yields only a marginal small improvement in the user's confidence in the information sought. The "best effort" principle in the search for information on the Internet is complemented by an iterative process that can be characterized as "good enough." This principle is akin to Voltaire's observation that: "better [than] is the enemy of good [enough]."[32] A "best effort" system is advantageous to users once they accept that there will always be an irreducible level of error, no matter the source. Consequently, they need to limit the number of sources of information consulted to the minimum compatible with this base level of error.

This new "good enough" paradigm is central to second life networks: for it is only by accepting a confidence level of less than 100% that users can effectively access the immense wealth of content available in networks. Perfectly certain information is rare, while practically certain information is infinite. Why deprive ourselves?

Clearly, the acceptable level of error is closely related to the kind of information sought. While user generated content is acceptable for the majority of the demands of information, it is evident that "good enough" is probably not a reasonable criteria for say, information regarding medical treatment or repairing nuclear power plan components or for understanding what airport runway signals signify.

The prolific advice available on the Internet can take many forms. One example are the direct opinions found on film sites either in the form of film reviews or simple tabulated frequencies of "Liked" and "Didn't Like" responses associated with a given film. This is also the case for food sites. It is also possible to obtain direct opinions or advice by posting questions on specialized websites.

Indirect forms of recommendations have also emerged. For example, suggestions can arrive in real time during a user's search, based on the context of the user's request. Suggestions can also appear from search engines as well as "collaborative filtering" algorithms. These latter methods correlate characteristics of a user's purchasing behavior with that of peers and make suggestions of the form: "users who purchased this product also purchased these others." Amazon first pioneered this approach for its online book sales.

In any case, reviews of content, products, and services by users are now omnipresent, whatever their form or method of dissemination. These bodies of reviews have increased in importance and influence as specialized websites have emerged and undertaken to codify, index and aggregate them.

Even if they can sometimes appear trivial and amateurish, user generated reviews and analyses can allow other users to discover things that they did not know existed. This can occur through the process of sequential verification described earlier where, through the abundance of sources and hyperlinks, new sites are stumbled on. This is another manifestation of the so-called "surfing" of the Internet. In this process of searching, the user can often add content to the sites visited in the form of comments or clarifications. It is here also that the role of the user as a node in the network is manifested.

In addition to abundance and redundancy of information, a third dimension allows users to navigate in the second life networks: reputation. When a user accumulates verifications on a given piece of information found on the Internet, it is evident that not all sources have the same relative value. The veracity we accord to information depends on the reputation of its source. It is highly likely, for example, that a 55-year-old executive will trust information found in The Wall Street Journal rather than one posted anonymously on a news blog. Inversely, a 25 year old looking for an opinion on Tim Burton's latest film would probably trust the opinions of friends on Facebook rather than those of a professional film critic on a film site.

Nevertheless, no centralized system exists today that rates users' "content-creating" reputation on the Internet. A reputation management system could be imagined that would become widely available and that would provide "trusted certificates" on request by users for content they are consulting. However, the extremely decentralized nature of the Internet does not bode well for the spontaneous emergence of such a service. Once again, there is no "President of the Internet."

Means of ensuring the reputation of content need to be found somewhere else, and on a case-by-case basis. For example, the open source collaborative encyclopedia Wikipedia is organized around different groups of experts and administrators, themselves Internet users, in charge of policing the site by, for example, removing obviously wrong information, and requesting clarification or "disambiguation" from the ranks of user-contributors. Even if this system increases the reputation of Wikipedia information, its quality of service is not guaranteed, since its panels of experts and administrators, largely unknown to most users, can make mistakes. Certain recent episodes have shown that it is possible for information to be modified, at least for a short time (including the biographies of living celebrities!).

However, even with such transparency, some level of vulnerability remains. There have been obvious attempts to change, delete, or add comments to articles. These have originated from individuals with different motives. For example, certain individuals quoted in articles have attempted to clarify, retract, or otherwise rectify what was reported. In response to this potentially harmful possibility, a student by the name of Virgil Griffith created a new tool called WikiScanner on August 14, 2007. WikiScanner allows the identification of the network or sub-network from which the edits were submitted. It cannot, however, identify the individual submitting the change nor if the organization from whose network the edit was uploaded was aware or even authorized the action.

According to Wikipedia, "WikiScanner also cannot distinguish edits made by authorized users of an organization's computers from edits made by unauthorized intruders, or by users of public-access computers that may be part of an organization's network." Griffith's WikiScanner FAQ (frequently asked questions) makes this qualification about any edit detected by the tool: "Technically, we don't know whether [the edit] came from an agent of that company; however, we do know that edit came from someone with access to their network." Although the FAQ goes on to say that "we can reasonably assume that any such edit was from an employee or authorized guest, there is no guarantee that the edit was made by an authorized user rather than an intruder..."[33]

One other simple way to verify an individual's or a site's reputation is to enter their name into a traditional search engine and count the number of times it appears. After all, few people have resisted the temptation to "google" their name.

"Googling"

Source: San Francisco Chronicle Parade

*

The second life of telecommunications networks is characterized by a fusion between physical networks consisting of copper wires, fiber, and routers, and social networks, consisting of people—ourselves. Practically speaking, their technical characteristics are fading in favor of their service aspects. This marks a successful milestone for the technological journey that began at the end of first life networks. Users today do not consciously feel they need to invoke physical networks as a precursor to accessing their social and professional ones.

If this new user experience represents a great leap forward for telecommunications, more than ever physical networks are the enablers. Similarly, physical and social networks will have to enhance and complement one another in order to evolve even further. This will have to come, on one hand, from the increasing transmission capacity and interactivity of physical networks, and on the other, from the improved software, intelligence and creativity of social networks. It is also important that the business models underlying the second life of networks be robust enough to ensure their long-term viability.

CHAPTER 4

Free as a Business Paradigm

The entry into the second life of networks is marked by disruptive technologies and usages, but also by disruptive business models. The telecommunications sector is seeing the emergence of a "free" service paradigm (or at least one that is perceived as free by end users) alongside the traditional "pay for use" models. These traditional models, ubiquitously used until the end of the last century, tightly coupled a monetary transaction to every unit of a service used. There was a sense of cost-causation. The individual causing a cost to be incurred through use of a scarce resource (that could therefore not be used by others), was directly responsible for its economic compensation. In the free model, the act of consumption and compensation are loosely, even distantly coupled.

Free is not a new economic concept. During the Middle Ages in Europe, the relationship between peasants and their lord was based on this idea of indirect compensation. The protection provided to the population by the local lord was part of a "non-merchandized" indentured relationship. Of course, the peasants had to perform mandatory duties for the lord and pay various taxes on their production. However, the taxes were levied as if they were disconnected, or independent from the protection ensured by the lord. The nexus between the payment to and the receipt of the Manor Lord's protection from external threat, natural disaster, or other danger were uncorrelated.

Since Jean-Baptiste Colbert, French minister of finance from 1665 to 1683 under King Louis XIV[1], what is considered as "free" are those things financed by the state, namely public services such as schools, health, as well as transportation infrastructures, e.g. roads, bridges and tunnels[2]. Again, the taxes levied to finance the state's treasury were uncorrelated with their spending.

More broadly, gift giving has always been an essential way to strengthen ancient and modern social structures. This is true within communities (for example to celebrate birthdays, marriage, and anniversaries). It is true for the knowledge transfer across family generations (for example the education of children, learning of language and teaching of first social reflexes). Finally, it represents the bedrock for collective creativity through the sharing of knowledge (as for example national public education, higher learning institutions, and government-funded research).

A number of North American Northwest Pacific Coast Indian tribal societies practiced a ritual ceremony, the "potlatch"[3], during which a family or leader of the tribe would host the event and each guest would receive a gift from the host in the form of objects or food. This had the objective of strengthening the relationships. These ritual donations were seen as free as there were no expectations of direct reciprocity. However, the donor (as well as the recipient of the gift) received prestige, honorary status and reinforced reputation, which compensated the cost of the potlatch. Some researchers have also attributed a wealth redistribution motive to the ritual.

Most of these historical examples have one aspect in common: the so-called gift, donation, or free service always results, in an eventual, albeit unconnected reciprocal compensation. The Middle Ages manorial lord and the 17th century Colbertian State both received taxes; the Northwest Pacific American Indian or friend who gives a gift had their self-esteem satisfied in certain ways. The desire to offer an appealing gift to the recipient, flatters the ego of the donor, as well as validates the effort it took to elicit the recipient's positive reaction.

In fact, bleak as it may seem, the existence of a genuinely unselfish motive for uncompensated giving, in all likelihood probably does not really exist *de facto*. Exceptions abound as in the obvious cases of family relationships or altruistically motivated organizations. The exchange in these cases cannot be characterized as economic but rather as genetic, psychological, or spiritual in nature.

Free, as used in common commercial practice today does not require the stringent conditions outlined earlier. Any initially free service or product whose compensation nexus is indirect, distant or generally uncorrelated in any obvious way to the initial free service or product will be regarded as "free" (or "pseudo-free").

The absence of a real or native concept of free is perhaps less surprising in the context of a free market economy, which for reasons of efficient equilibrium, requires that a transaction compensates the resources implicit in the gifted product or service. By contrast, with the dawn of the twentieth century, our modern economies have seen the emergence of the first manifestations of the weak form of free.

In particular, one of the first examples of a marketing method based on free was that of King Camp Gillette's: razors were free or low cost, but they consumed the more expensive blades (necessarily so to cover the total cost of the two elements and to generate a profit). Other products have used the same pricing model. Video game consoles are sold at subsidized prices in order to encourage customers to buy high margin games. Inkjet printers are priced as loss leaders to generate consumption of the ink cartridges sold at a substantial mark up.

The Free Business Paradigm and Telecommunications

Telecommunications and its adjacent sectors, such as microcomputing and software, are where the free pricing model has been most closely associated to their development. The oldest example is the phone set used with fixed telephone lines: it was part of the subscription and in that respect was free to the users.

The Free Business Paradigm in Telecommunications
Source: San Francisco Chronicle Parade

Later, in the competitive mobile phone space, operators began to develop creative commercial packages to attract potential customers. These offers included low prices for the mobile handset, sometimes very close to zero, bundled with a subscription commitment of several months. What this did is shift the payment nexus for the handset several months into the future. Customers tacitly understood and approved the fact they would be paying over the length of their subscription what they would have had to otherwise pay upfront.

This is the case with monthly subscriptions as well as with pay-as-you-go or pre-paid minutes of use plans. This arrangement had positive results in the sense that it lowered the upfront cost barrier for the majority of users: mobile phones can be too expensive for many people and this formula made the mobile experience more affordable and accessible. This "win-win" form of credit helped make the mobile telecommunications market more fluid and, in doing so, accelerated its rate of penetration.

The deregulation of the fixed telecommunications sector, which began in the late 1980's, transformed the global telecom value chain. It also allowed the free model to develop further. This "new deal" in telecommunications, which saw the emergence of so-called "alternative" carriers, led to price wars as a way to acquire and retain customers. Indeed, the reality was that these new players could only rely on price differentiation as a way to build their new businesses. This is because they did not possess at first infrastructures of their own upon which they could build differentiated services.

When low-speed, dial-up access to the Internet was first introduced, the price war manifested itself in the form of a "free" access to the internet for only the cost of a local phone call. This model exploited a regulatory ambiguity in the status of the Internet regarding whether it was subject to the same regulations regarding tariffs and revenue sharing as telecoms network operators. This phenomenon first appeared in the UK in 1998 with the company Freeserve, an internet service provider (ISP). Freeserve, which was subsequently acquired by Wanadoo, a subsidiary of France Telecom, offered its customers free Internet access. The underlying economics that made this work was that Freeserve allied itself with the competitive carrier Energis and collected revenues from British Telecom in the form of revenue sharing fees. Since Freeserve and Energis were in fact generating new traffic on the BT local network, traffic that BT would not otherwise get, it provided a compensation fee to the originator of this extra revenue-generating traffic. This model spread to several other countries where these revenue sharing fees were sufficiently high to make the system work. However, in early 2000, this free internet access model was made obsolete by the deployment of high-speed access networks such as ADSL and cable-modem, as well as reductions in revenue sharing rates which made them less attractive. In the United States, in 1998, the company Netzero, an Internet service provider that came out of the incubator Idealab, pursued a different model for providing free Internet access to its users. Netzero used an advertising-supported model which consisted of displaying so-called "banner ads" on users' PC screens. The resulting revenues compensated the cost of the local call. In these first two "free" models of the Internet, the increased competitive pressure on the use of the network resulted in a shift of revenues. Specifically, those associated with the B2C, or "Business to Consumer" model (i.e., where services flow from producer to consumer and revenues flow from consumer to producer), shifted to a B2B, or "Business to Business" model (i.e., where services still flow to consumer but revenues flow from producer to producer). In the first case revenues consisted of the revenue sharing fees and in the second case of advertising fees.

With the rapid development of broadband Internet, the network operators began to compete through packaged offers. These offers were in fact a new stage in the emergence of the free, or more precisely "quasi-free," paradigm in the world of telecommunications. In a sense, it has profoundly changed the mode of consumption from one where the customer paid in proportion to use, to one where usage was "unlimited." The packaged offer model illustrates another facet of the concept of free: it is not the shifting of a source of income to a nexus in the value chain less visible to the end customer, but rather a simplification. This is achieved by leveling the rates in a way that allows in theory infinite use of the service (obviously impossible in practice) for a finite price. In theory, "at the limit," (or, as mathematicians say, as use approaches infinity), a minute of Internet use approaches a zero price asymptotically.

More fundamentally, from a conceptual and maybe even philosophical perspective, the Internet has massively imported to telecommunications networks the ideal of free. From its inception, the Internet was conceived above all else, as an open space, available to all at no cost. More specifically, the Internet is at the origin of the notion of openness adopted by networks: that everyone should have access to the information, content, and applications circulating on the Internet, and have the potential to modify and augment them. Later the principle of openness gradually came to be confused with a very different principle of free (what is free is not necessarily open and *vice versa*). Originally developed for the military, the Internet quickly became a working tool of scientific research. In this environment, researchers were accustomed to share their work with peers around the world. This free flow of information goods is a basic, tacitly held principle of the scientific community. Scientists are used to communicating widely and freely in conferences and publications. Their interest is in fact, to publish their work in order to improve their reputation and peer-recognition: potential monetization of their work is not their primary concern.

Today, despite the fact that the Internet has lost its purely academic focus, it has retained its original researcher's, scholar's, and user's principles of community exchange, collaboration, and knowledge sharing. Therefore, any Internet user can access and contribute to the communally developed knowledge and content available and in doing so, generally receive more than they contribute.

This return to community, to looking for "contacts" through networks, in today's individualistic and introspective market economy dominated society is remarkable. Despite the seeming paradox, this grass roots movement is flourishing even though it is rooted in capitalistic topsoil. This "techno-communalism,"[4] results from reinforcing trends in capitalism, communalism, and technology. All these strands are brought together in the motto of new Silicon Valley entrepreneurs: "make money, change the world."

This tradition of openness, community and, free access—if one can already speak of tradition in the relatively nascent Internet space—is in fact propagating. Early conceptions of the networks of the future include the idea of "Cloud Computing" which proposes that computer intelligence be fragmented and distributed across networks. Today, this intelligence is mostly concentrated in the server farms of Google, Yahoo, Microsoft, Amazon, eBay, and IBM.[5] This will facilitate the spread of the principles of free access to resources and remixing of information on the networks that constitute the Internet[6].

Open Source: the "Free" Model Comes to Software Distribution

As Internet brought the paradigm of "free" into telecommunications, the "open-source" model is introducing "free" into software on a large-scale basis. Unlike proprietary software, for which customers must pay commercial use licenses, open source software is literally software whose source code is made available for free. The ability to modify, re-mix, improve, or expand the code by adding complementary routines and then redistributing it back into the user community which can in turn launch another cycle of improvements is probably the most important aspect of open source.

The history of the emergence of open-source software is replete with instances showing the reluctance of the established players to join in the surging adoption of the open source principle.

David F. Marquardt, a Microsoft board member during the 1990s described[7] the challenge that the "free" model represented for Microsoft. "He recalls that he was 'amazed' that, [despite Microsoft's characteristic closeness to its markets and customers], it was putting so little into the Net. ...'They weren't in Silicon Valley. When you're here, you feel it all around you,' said Marquardt, then a general partner at Technology Venture Investors. He broached the subject at the Microsoft April board meeting in 1995. He goes on to recount Bill Gates' response: 'His view was the Internet was free,' says Marquardt. 'There's no money to be made there. Why is that an interesting business?' "

On June 1, 1995, 40 Microsoft executives gathered at the Red Lion Inn in Bellevue, Wash., to brainstorm an Internet strategy. Bill Gates gave a 20-minute talk on the "Internet Tidal Wave." "[Microsoft project leader for Internet Explorer Benjamin] Slivka's scheduled 15-minute talk ended up lasting more than an hour. 'I got some people riled up,' he says. At one point, Slivka proposed that Microsoft give away some software on the Internet, as Netscape was doing. Bill Gates, he recalls, 'called me a communist'."[8]

"First, we need to discredit the public's notion that the best things in life are free."

Source: Harvard Business Review

Initial reluctance aside, the open source movement was key in driving the largest PC software publishing companies like Netscape, Adobe, Macromedia,[9] and Microsoft to give away free products such as the PDF reader, the various Internet browsers, software updates and bug fixes, as well as "drivers" for peripherals (printers, webcam, BIOS, scanners, music players, and other hardware).

More recently, the services and content market sectors have had to deal with issues associated with the "free" paradigm. These sectors are still in the process of finding the best compromise between covering costs, growing their customer base, and optimizing profitability.

In the older models of online services like the premium rate telephone service dial-a joke[10] there was no monthly subscription paid in advance; it was a pay as you go model. In the first generation of commercial online services like CompuServe and AOL users had to pay a subscription with monthly fees. There were also some extra fees for "premium" content for which third party content providers were remunerated when their services were used.

None of these types of model for premium content survived transplantation to the Internet. Newspaper and magazine websites have tried with only relative and limited success to charge directly for their content through flat-rate monthly subscription or at least to get users to pay for premium services like archive access. This type of pricing represents a special type of subscription, referred to more generally as a "freemium" combining "free" and "premium" elements. The user has free access to part of the content but pays a monthly premium for extended access and sometimes, additional functionalities such as increased storage space. Although there have been a few attempts by web sites in the past to deploy a micropayment system to collect occasional, non-subscription payments; today it is rare to find a site that restricts access to their content or conditions content access by a prepayment.

Free Internet e-mail services like Microsoft's Hotmail later unsuccessfully tried to charge their customers either for the quantity of messages that could be stored on the system, or for using special messaging software other than the Internet browser. Google launched its own e-mail service "Gmail" in April 2004, which interdicted other offers by providing users not only more storage space (one gigabyte) but also by making it free. All of Gmail's competitors subsequently began offering users enhanced services at the same "free" terms including equivalent storage space.

In May 2007, Yahoo responded by offering unlimited storage on its web based email service, Yahoo mail. The service had at the time over 250 million global users with "normal email practices," that is, not engaged in activities like using Yahoo mail for basic online storage[11]. As Chris Anderson noted in his article in Wired Magazine on the free economy *Free! Why $0.00 Is the Future of Business*: "So the market price of online storage, at least for email, has now fallen to zero. And the stunning thing is that nobody was surprised; many had assumed infinite free storage was already the case."[12]

From the start, for a number of reasons, players in the upper layers of the networks value chain, particularly those involved in Internet services, faced difficulties in charging users directly for service usage as a way to ensure their profitability.

Certain player's offers were amenable to "full collaborative" development through their users' contributions. This is the case of Wikipedia, the popular online encyclopedia launched in 2001, where Internet volunteers write the articles found on the site without financial compensation.

Could it be that, with Wikipedia, the Holy Grail of the "real" free model has been found at last, where the act of giving awaits nothing in return? In fact, even though writers of articles receive no real, tangible compensation, their motivation may be rooted in the satisfaction of seeing their own knowledge and work published online, on this globally popular site. Similarly, as is the case with traditional gifting, the resulting self-esteem, sometimes bordering on narcissism, can indeed be a powerful motivation driving Internet participation. Nevertheless, one of the original founders of Wikipedia Jimmy Donal "Jimbo" Wales (http://en.wikipedia.org/wiki/Jimbo_Wales), seeing the commercial potential of transposing the Wikipedia model to social search created a private company called Wikia (http://en.wikipedia.org/wiki/Wikia). The emergence of social networks these last few years has only accentuated the importance of this type of service.

Indicative of the times, the most authoritative paper based encyclopedia, Britannica decided in April 2008 to become available online to bloggers, online journalists, and web publishers *via* Britannica Online (www.britannica.com) for free[13].

The company also still publishes the 32-volume Encyclopaedia Britannica. "It's good business for us and a benefit to people who publish on the Net," said Britannica president Jorge Cauz. "The level of professionalism among Web publishers has really improved, and we want to recognize that by giving access to the people who are shaping the conversations about the issues of the day. Britannica belongs in the middle of those conversations." He goes on to say: "Bloggers, journalists, and Web sites link all the time, of course, but they may not realize they have the option of pointing to Britannica articles. So let me be clear: they do." He adds "... Web publishers can link to as many Britannica articles as they like[14]."

Another kind of "collaborative" network is those called "peer to peer" or P2P. Peer-to-peer networks have spontaneously emerged and flourished around the Internet. Their purpose was to facilitate direct exchange of a wide range of content between users. The first of these networks were designed to share music and were represented by companies such as Sharman Networks with its Kazaa software, or open source applications like Napster, eDonkey2000 and eMule. With these services' software, users make available their computer resident music libraries to other users. In concept, a given music library could be replicated across the Internet *ad infinitum* online (often, but not always, illegally). Eventually other content was also exchanged across peer-to-peer networks, including computer software programs, video, and voice, the latter over established services like Skype, which allowed for free phone calls between computers. Skype users download onto their computers the Skype software. In fact, and this is little know by users, this software installs a computerized internet call server, so that the Skype users community becomes a node in a voice communications subnetwork connected to the Internet.

In this example of the peer-to-peer network model, the free calls enjoyed by users are compensated by contributing their PC to the Skype network and agreeing to let it install the call server software and control its routing functions. Therefore, in reality the users cede control of a part of their PC to make their contribution to the Skype peer-to-peer network.

In all cases of collaborative networks, the free system is completely decentralized and compensation, even when it is abstract (like peer recognition), takes place between users. Virtually no costs are borne by a central structure since there is no source of directly tangible financial revenues in these models to justify a large central organization.

However, in the other cases, a commercial company provides the service and expects tangible and direct compensation. This is the case of search engines, e-commerce, social networking, and photo-sharing sites. These environments, however, are not conducive to up-front billing of users. The financial and economic realities of these companies soon led to the development of new, creative, and profitable business models requiring no direct compensation from users.

The Advent of the Online Advertising-Supported Model

The first online services guides and directories offered by Internet search engine providers like Google and Yahoo! also became the first instances of having a service's use compensated by someone other than its primary user. This has always been the case previously with printed phone books, which were generally provided free. Some companies like Yahoo! and Excite, combined content and services to "retain" visitors on their web sites (their "portals") and sold the space available for ads on their web pages to advertisers interested in reaching these visitors.

Therefore, from the Internet service provider point of view, revenues do not come directly from users but from advertisers. From the users' perspective, they do not pay directly the service, which they perceive as free. Rather, they indirectly pay for the cost of the ads when they buy the products and services of the company placing these ads. For example, the user who reads a media company's newspaper online can access and read many articles without paying for them (with the exception of archives in some cases). If this user buys a car of a company that advertised on this newspaper site, then a part of the price of the car compensates the advertising spending.

The advertising supported model has grown exponentially with networks' entry into their second lives and the accompanied expansion in the number of content and service sites. This model had already been widely used in the context of other communication networks and media.

Commercial television and radio understood early on the virtuous circle of the advertising model. On the one hand, advertising finances and facilitates access to TV and radio stations' programs by shifting the financing of its production and distribution to businesses wishing to market consumer products and services. Otherwise, the cumulative amount of investment needed would be a significant barrier to entry and would certainly result in much less content production. On the other hand, free access to a large number of TV programs increases their audience, which is what advertisers and marketers are looking for and are therefore willing to finance the programs that create these audiences.

However, these pioneers in the use of advertising in radio and TV, have nevertheless, reached in recent years the limits of full dependency on advertising revenues (excluding access fees in the case of cable TV and Satellite TV). First, in addition to publicly available television financed by advertising, pay TV emerged as a complementary option for accessing more diversified content (movies, sports, news, ethnic language programming and music), even though it did not completely eliminate its dependence on advertising revenues.

Second, cable TV and satellite TV used new technologies that allowed a significant increase in the number of programs and content by providing a much larger distribution capacity than Over-the-Air TV could.

The success of these new Pay-TV initiatives encouraged the telecommunications operators themselves to invest in the production, co-production, and distribution of new programs.

Third, new initiatives were launched to offer special services to bypass the ad messages embedded in TV programs by personal video recorders and associated services such as TiVo. This company created in 1977 designed and marketed a digital PVR (personal video recorder) which among other creative features had a commercial-skip feature in addition to time-shifting features. This system, as of January 2008[15] was used by about 4 million subscribers mainly in the US, is now part of several TV set-top boxes.

What is ironic, however, is the fact that TiVo has since changed its attitude significantly *vis-à-vis* advertisers and their budgets. "Now, TiVo is helping networks and advertisers overcome that technology by offering interactive banner ads that make sponsors' names visible as their ads are being fast-forwarded. And its ability to keep second-by-second ratings on the commercials that, after all, some viewers do still watch, also helps."[16]

The completely ad-supported free daily newspapers are a more recent manifestation of the free business model that is likely to continue expanding worldwide.

Several initiatives were launched in the past to introduce advertising in "pre-Internet" online systems such as videotex, most notably in Europe. For example in France for the electronic phone directory available on Minitel videotex terminals, the first two minutes of use were made free as a way to grow audience. This was analogous to the freely distributed paper directories available at the time. Other examples include the Alex-Telidon system from Bell Canada, AT&T's Viewtron in Florida, Gateway from the Times Mirror newspaper group in Los Angeles started in 1986, and Prodigy from IBM, CBS, and Sears Roebuck in 1990. Indeed, in 1980, while the first systems of videotex in Europe (the Minitel in France, Prestel in the UK and BTX in Germany) were fighting to impose their technical

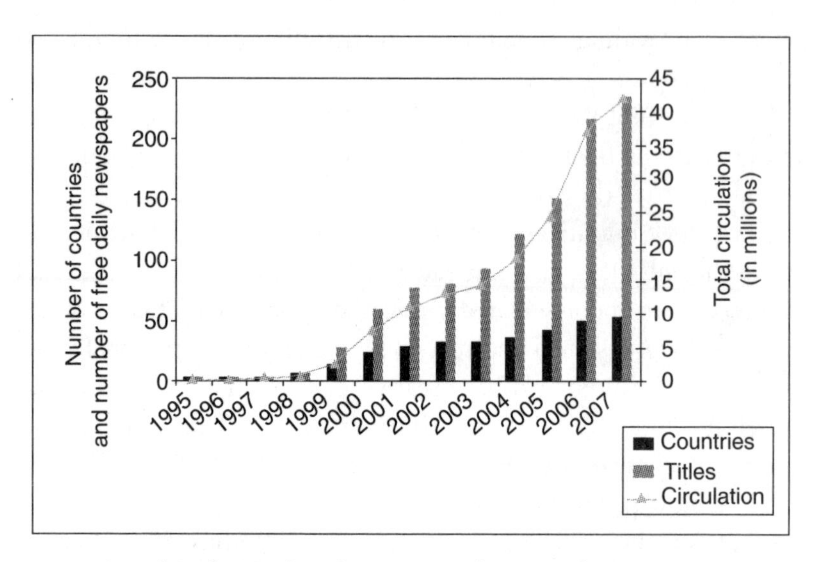

Growth in the Number of Countries with Free Daily Newspapers, their Number and Total Circulation

Source: http://www.newspaperinnovation.com/

standards around the globe, the North American videotex industry, led initially by Canadian companies justified the need for more sophisticated color graphics so as to display high-quality advertisements and draw advertisers.

The Internet quickly became the network of choice for developing new creative applications and services. This was due to in part to its simple technical standards, which made it easily accessible to entrepreneurs as well as to its global coverage, which made any product deployed upon it available to a large number of users. Therefore, the very first appearance of large-scale advertising on a network, in this case the Internet, goes back to the mid 1990s, in the wake of the emergence of Netscape's Mosaic, the first commercial Internet browser.

The company HotWired, the first magazine on the Internet, in its search for additional funding beyond the model of sponsorship by institutional advertisers, was led to invent the principle of "banner advertising," which would display so called ad banners on one

The first page of the HotWired website

Source: http://www.flickr.com/photos/veen/188264269/

or more pages on the site. AT&T is often cited as the first advertiser on HotWired to use a "banner" ad.

The development of the advertising model on networks, though it may seem insignificant in a world where many manifestations of the free paradigm already existed, constituted the vanguard of the "free" revolution in the second life of networks.

The simplest manifestation of online advertising remains the "banner" which appears on the side of the screen (or sometimes even in the middle or above), randomly, on which one can click to go directly to the site advertised. The payment of the site that

First banner ad on a web site in 1994

Source: http://thelongestlistofthelongeststuffatthelongestdomainnameatlonglast.com/first66.html hosts the advertising varies first in relation to its audience, then depending on the number of effective clicks by users. For example in 2008 for the US, the display of banners on a site with an average number of visitors, but for which there is no "click-through," cost the advertiser a few dollars per thousand exposures (the "cost per thousand" model). In the case where the advertiser pays only if the user "clicks" on the banners, the "cost per click" model will cost more than 10 dollars per thousand clicks. Finally, if the advertiser is charged only on the basis of actual purchases made by Internet users on its site, the "cost per action" can exceed these amounts, depending on the price of the products and services purchased.

Over the last few years, many technological improvements have enhanced the advertising model. Google's initial innovation was to link the context of the simple text only advertisements shown to users to the words they entered into engine's search box. Google sells certain keywords to advertisers, which trigger specific correlated ads when these keywords are invoked. These ads are then highlighted in the upper right of the screen, alongside the search results. This direct link between the instantaneous interest of an individual and the context of the advertising message received was considered a major innovation at the time. Today it has become the core competency of search engine companies and Google, more than the others, benefits the most. This system is always being improved and adapted, in particularly to new services such as mobile telephones and Internet-mediated TV.

A second innovation was the syndication of advertising associated with search engines like Google onto sites or blogs that want to take advantage of revenues derived from advertising but do not want to become advertising brokers. A good part of today's financing of new Internet web companies and sites is due to this syndicated advertising.

Finally, a series of innovations linked to progress in context and user profiling algorithms is improving the effectiveness of the delivered advertisements. For users, this will represent a convenience, one that will transform their perception of ads from an intrusion into their internet experience to an informational service that will enrich the experience. The idea is not only to link the ad to the current activity of the user (such as the site being visited, and the subjects searched, for example), but also to their more permanent tastes and preferences derived from personal data recorded in conjunction with site visits (assuming the user has chosen to allow such transparency). Such a system, assuming it can maintain strict privacy protection, involves a level of behavioral analysis of users that can result in levels of relevance for both users and advertisers that approaches prescience.

This "context" is in fact available both in space and in time. Advertising that accounts for a users' geographical context is beginning to appear on mobile phones. For example, a user may get a message that there is a highly recommended Italian restaurant nearby if their profile indicated a preference for this type of cuisine. This type of location-based profiling is possible because of the growing availability of Global Positioning System (GPS) chips embedded into most mobile telephones today that measure the geographic position of the user[17].

The cost of different forms of online advertising depends here again on their commercial value (i.e., the probability that the user clicking on an ad will result in an actual purchase). A customer, who enters "flights between New York and Paris" on Google and then clicks on the first sponsored travel agency link, has a high probability of making a purchase. This probability is higher than the one associated with clicking on an advertising banner that promotes flights from New York to Paris. Factoring in the "context" of the user can vary by a factor of two the cost per thousand paid by the advertiser (but of course there is a corresponding increase in the purchase conversion rate).

The search engine companies like Google, Yahoo, and MSN continue to dominate the field of online advertising through continuing innovations in profiling. Previously, the analysis of the behavior of users and their targeting was mainly conducted on the larger sites, including multi-services portals, which as a whole represent the lion's share of traffic. This analysis, however, was more basic, less automated, and relied less on statistics regarding site visits and search queries made by users. A study by the New York company Avenue A/Razorfish in October 2007 found that 54% of Internet users start their shopping on the Internet *via* a search engine rather than going directly to brand-controlled sites.

Worldwide, the relative weight of Internet advertising spending compared to that of other media is still relatively low (about $40 billion worldwide in 2007). In the United States, the most dynamic market for Internet advertising, it still only accounted less than 10% of all advertising spending in 2007. However, Internet advertising has exhibited the largest growth for several years. In 2007, advertising on the Internet had already surpassed radio advertising, for example.[18]

Social Networks and Advertising

Social networks like MySpace and Facebook can naturally bring a new dimension to online advertising with their ability to attract loyal audiences who are often permanently connected, either directly *via* PC, or increasingly *via* their mobile telephones. More importantly, social networks can allow marketers to show better "profiled" and therefore more efficient ads to their audiences. In November 2007, Facebook launched SocialAds, its proprietary digital advertising program. In this system, advertisers associate ads with specific desired profiles of customers (sex, age, nationality, education, etc.). The SocialAds algorithm will then deliver the ad to those Facebook users matching the associated profile selected from among its some 70 million regular participants[19].

Social networks can also offer other business opportunities for advertisers. By their nature, social networks develop links between their participants based on their affinities. These links reveal, in indirect and unconscious ways, information regarding the tastes and preferences that motivated the individual users to join a given interest group in the first place. As a result, the knowledge of the attributes of any user opens the door to those of dozens or hundreds of their contacts who can be expected to similarly reveal and share at least some of their interests.

Facebook also launched its "Beacon" system in November 2007, in which any purchase by a member on a Facebook partner site, such as Blockbuster or Conde Nast, triggers an immediate advertisement from this site to all of this member's contacts or "friends." Therefore, even actions taken by a user on sites outside Facebook, are shared with their social network inside Facebook.

Beacon caused a major outcry among Facebook members, some of whom signed an online petition against its use. In the end, the founder and CEO of Facebook publicly apologized and introduced a system where users had to explicitly agree or "opt in" in order for the Beacon functionality to be activated. Previously, it was activated implicitly without any specific consent of the users beyond agreement with the general terms of use of the Facebook service. This presages a growing need for the companies using the marketing capabilities enabled by the new profiling technologies to address privacy and personal data protection. Any deficiency in this sensitive and dynamic domain could, not only potentially harm individuals, but also collectively deprive users of what are arguably useful and relevant services. In March 2008, in response to the security concerns of users, Facebook said it would also implement extensive privacy controls.

In quantitative terms, advertising on social networks is growing at a rate three to four times faster than that of overall Internet advertising. According to the market research firm eMarketer, while online advertising expenditures as a whole rose by around 20% annually, ad spending in social networks rose from 1.2 billion in 2007 to 2.1 billion in 2008, representing a 75% increase and is expected to reach $4 billion in 2011 (for a market of over 100 million users of early social networks). In the United States, advertising revenues generated on social networks are expected to grow by 70%, from $920 million in 2007 to $1.6 billion. However, there are already signs of relative customer fatigue in social networks[20].

Content and Advertising

The advertising-supported model has gradually infused all forms of online activities and commerce, particularly on the Internet. Virtually any digital product or service became available for free online. This is particularly true of all digital content, and in particular, music, which has been the subject of ongoing legal battles between the music industry and the Internet peer-to-peer music file sharing services[21].

Music

The record labels Universal Music Group and EMI entered into an agreement in 2006 with SpiralFrog a web site that distributes their music for free online using an ad-supported model based essentially on banner ads. EMI Music even announced in January 2007 the signing of an agreement with Chinese search engine Baidu, the first to launch an ad supported music distribution service over the Internet in China.

Ad-supported music is also growing on social networking sites. In October 2007, EMI entered into a contractual agreement with IMEEM, a social network to make its music catalog available online *via* ad supported distribution.

Sony BMG and Warner Music Group had also previously signed the same kind of agreement. The largest social network MySpace launched in 2007 a creatively implemented free music offer. If users accept a brand as a "friend," then they are rewarded with free music downloads.

In the same vein, Napster, a pioneer in the internet peer-to-peer file exchange (especially music), made a return to the (legal) Internet music scene with a free ad supported music service[22]. In any case, whether they are forced by law or because of fear of lawsuits, these music sites pay some sort of fees (not necessarily proportional to the advertising revenue actually generated) for the rights to broadcast and distribute music online. These payments are made to one of several rights management companies including BMI (Broadcast Music Incorporated), ASCAP (The American Society of Composers, Authors, and Publishers), and SESAC (Society of European Stage Authors & Composers).[23]

Free models other than advertising-supported ones have also appeared in the online music field. The commercial initiative of Universal Music in late 2007 called "Total Music" introduced one such new model for the marketing of music on mobile phones. In this case, the manufacturers of cell phones or the wireless operators assume the subscription cost of about $5 per month. This allows for an unlimited music service from "Total Music," for which the end-user pays nothing directly for the service and thus perceives it as "free." In principle, this system is similar to the free model existing in the wireless sector where carriers subsidize handsets and appear to users as almost free.

Nokia, for its part, introduced in December 2007 a new service called "Comes With Music," in which Universal Music Group International makes available on the Internet and mobile networks its entire catalog of music and songs for unlimited consumption for a year, to every purchaser of a Nokia mobile phone. The user does not pay anything for the music itself and Universal as well as recently joined EMI receive a one-time payment of $80 per compatible handset for a one-year unlimited download from their respective music catalogues.[24]

What is surprising is that along with this proliferation of free models for online music distribution, premium pricing models radically different from the free models are also thriving. For example, some types of consumers—particularly those of Generation Y—are willing to pay more for a 30 seconds ring tone than what the complete music track from which it derived cost.

Video

As in the case of music, the distribution of online video has also become largely free. The distribution of short-form video or clips, popularized by sites like YouTube and Dailymotion were initially completely free, without advertisement, and as in the case of music, posed similar legal problems. The advertising model was subsequently introduced as a way of funding these services' operations. In the case of video, a richer set of options for inserting ads was available in addition to traditional banner ads. These included direct insertion before, during, and after the videos (known as pre-roll, mid-roll, and post-roll video advertisements, respectively). Classic banner ads are also widely used to monetize freely distributed videos, and have begun in 2008 to appear on mobile phones.

Software

The world of commercial software could not resist the power of advertising. A number of Internet sites have emerged providing free software (freeware) compensated by the associated advertising revenues.

Even if the freeware model was in use before the Internet came into prominence, free software gained momentum once it was provided as a service available on Internet sites. Google was a pioneer with its search service and later with its free office applications suite "Google Apps," both are supported by advertisements, which in the latter, can be avoided by subscribing to a for-fee version. Microsoft similarly provided in addition to a search engine, "Office Live," a set of similar Internet-resident, advertising-supported applications and announced in the summer of 2007 the release of a free version of its product "Works" also financed by ads, but this time as a download on users' computers (the ads are updated upon each (required) connection to the Internet). Adobe also launched a free online version of its flagship image management software "Photoshop," financed by advertising.

The significance of the free, advertising supported model is far-reaching: it is, for example, a plausible model for distributing enter-

Source: http://www.blaugh.com

prise software. Vauhini Vara comments in his article *Companies Tolerate Ads to Get Free Software*: "Consumers have been doing this for years, using free online software services like Yahoo Inc.'s email site or MapQuest Inc.'s maps, and, in exchange, putting up with the ads that run alongside. Conventional wisdom had it that corporate users were not willing to see ads in their online software."[25]

The company Spiceworks provides SME's (Small and Medium Enterprises) free, advertising supported software to manage networks and information systems. This company came up with the idea of developing a simple product, inspired in part by Apple's iTunes service for music and video downloads[26].

Another example of a company designing software for professionals funded by advertising is Practice Fusion, which allows physicians in the United States to store and retrieve the records of their patients free of charge. Electronic patient record management systems can otherwise cost up to $20,000 to purchase[27].

A survey by McKinsey & Co. / Sand Hill Group found that businesses are increasingly accepting of the ad-supported software model. Their propensity to test new business models for software acquisition is greater today than during the first period of the Internet from 1994 to 2004[28].

Games

Video games are another type of content becoming increasingly available as a free ad-supported product. In this area, the creative concept of inserting "ads" within games has emerged[29]. The company NeoEdge, for example, is active in the business of delivering simple ad supported free video games on the Internet. Kongregate.com is a sort of "YouTube" of video games, is in fact a social network allowing developers to create ad-supported games in both single—and multi-player modes. Not all the players in this new market are newcomers: Microsoft acquired in 2006 Massive Inc., a company specializing in the placement of advertising in video games. According to Yankee Group One, revenues generated by ads inserted in video games are expected to reach $970 million in 2011 up from a relatively modest $80 million in 2008.

Telephone

In a kind of historical replay, the ad-supported model having fully invested the Internet, is beginning to insinuate itself into telephony. Free PC-to-PC voice communications is today an established feature of the Internet; provided of course, that its users pay for Internet access. Recently however, PC-to-fixed telephone and PC-to-mobile telephone calls, hitherto provided on a paid basis, began being provided on an ad-supported basis by companies such as Jajah and Talkster. However, if these communications are free, the persons involved in the call must meet certain requirements (such as being in particular geographic areas), which reduces somewhat the scope of these creative initiatives.

Another company in the United States, Pudding Media, has been working on an alternative, but rather intrusive model. Users must accept that their conversations be registered by an A.I. (artificial intelligence) module that uses sophisticated voice recognition algorithms to identify key words spoken during the conversation. Based on this information, the system will select and display relevant ads on the user's computer screen during the course of the call. This limits the service to only calls made from a computer.

An initiative launched in the United Kingdom in 2007 in the wireless space by the company Blyk provides free mobile communications service to consumers supported by advertising. This service targets young people between the ages of 16 to 24 willing to provide their user profile. They receive *via* SMS short advertising text messages, correlated to their profiles. This suggests that users may have a higher propensity to accept and watch ads on their mobile phones (especially during down-times such as waiting in line and the like).

Telephone sets themselves have also become the targets of advertising. In 2007, Google launched its mobile phone, framework nicknamed "Googlephone" or the Gphone; which consists essentially of a mobile device operating system and additional software ("Androïd") The technical specifications are openly available and royalty-free. The devices are intended to be built by manufacturers throughout the world. While it is too early to tell, it will most probably be financed by ads displayed on the screen, which could also contribute to financing in part or in whole the network usage.

The success of the mobile phone designed by Apple, the iPhone, is due, in large part to its design and compelling services. These two factors have allowed an unmatched user experience that includes one of the most ergonomically superior Internet navigation approaches available today. The iPhone is said to have been responsible in January 2008 for 0.13% of all pages viewed on the Web. This is a considerable proportion for a mobile phone, which, moreover, was launched only a scant six months before in the United States.

Finally, in the same spirit as the ad-supported free service model, free Wi-Fi was introduced. An example of this is the launch by CBS in New York November 2007 of the "CBS Mobile Zone," which consisted of twenty, ad-supported "hot-spots" (Wi-Fi access points). The Sunnyvale, California company AnchorFree developed an electronic ad distribution service to support deployment of free to use Wi-Fi networks, especially in hotels. However, doubts remain about the viability of this economic model, because the ads are generally displayed on the home page of the Wi-Fi network, which appears to be too restrictive and for the most part insufficiently adapted to the needs of users to represent an effective advertising vehicle.

The Sustainability of Advertising Models?

At the beginning of the second life of the networks, while the network infrastructure continues to be financed exclusively by its customers, increasingly through packaged flat fee payment models, use of new services is financed by advertising models based on audience—that is, by the traffic carried over these infrastructures. It is essentially only the new service and content providers including content aggregators such as the television and radio networks, which have managed so far to take advantage of the momentous technological innovations we have seen. This has been made possible by the creativity of these players and the increasingly available physical distribution and transmission capacity, without which the traffic generated by these new activities could not flow.

Experience shows that the only stable "free" system is one that results in a "zero-sum game." There always is, in one form or another, a counterpart to a transaction described as "free" of charge. Any revenue that is shifted in time or in space, ultimately must return to the player who has borne the costs of providing these "free" goods or services (such as subsidized terminals). However, when the system "leaks" and does not completely compensate, in due proportion, the players bearing its costs, the system can become unstable.

Is there a belief today that the advertising supported model is not viable? The answer is no, if we are to judge it by its exceptional widespread use and deployment in a large number of services and content. This paradox can be put into perspective by looking at the historical context in which these new services developed. Their emergence late in the 1990s during networks' first life allowed them to benefit from two factors. They benefited from the almost complete depreciation of these networks' copper-based infrastructure as well as the over-investment in infrastructure that occurred during the Internet "bubble." Consumers gained tremendous advantages from the seeming "windfall" represented by free services. These advantages, however, did not simply spring up out of air and at no cost.

Within the value chain, not all players, even today, seem to have benefited from or adopted the advertising supported model. This is the case for artists, musicians, actors and more generally, content producers. In contrast with network operators, these players have not fully "amortized" their "investments".

The parallel between the network operators and content producers is actually quite profound. Both must initially invest massively in order to provide their services. Network operators had to invest in copper wires, fiber optics, routers, antennas, as well as points of sales, technical assistance, and sales support, and call centers. Content producers must develop human capital in the form of ideas and intellectual or artistic creativity. The fixed costs associated with all these investments are high.

Both of these types of investments also have low variable costs. Once content is created or digitized, the cost of distributing is almost zero. Once copper or fiber access is installed and connects users' devices and home computer has access to networks worldwide, it technically costs almost nothing for the user to browse the Internet or to download content. In economic terms, the marginal cost of navigating the Web and consuming content online is almost zero while their fixed costs are massive.

This is perhaps the crux of the explanation of the original leakage of advertising revenues from the network and content layers to the services layer of the telecommunications value chain. Indeed, if "it does not cost anything extra" for networks to carry packets of bits of a digitized song, why would the singer and the networks need to be paid? But this "collective amnesia" of the value chain, which forgets that some players had to invest so that these particular arrangements of bits can exist and be shipped quickly, reliably, to whomever is interested in receiving them, creates a palpable risk that these participants may no longer invest at all, or as much.

This collective amnesia reflects the fact that the pathways carrying the signals required to resolve the "zero-sum" game described earlier are compromised. Repairing these pathways is central to ensuring the long-term viability of the current arrangements describing the telecommunications value chain.

At a more macroeconomic level, the development of the free model on networks also raises questions, which, if they worry some analysts, are also economically stimulating the entire telecommunications sector. The ad-supported business model dominating the Internet has also led to a financial transfer from Europe and Asia to the USA.

There is a growing financial transfer to United States based online advertising companies. For example, if we consider that American players captured about 70% of on-line European advertising revenues in 2007, the corresponding transfer of wealth from Europe to the United States was about $7 billion. Similarly, the transfer of wealth from Asia to the United States is already estimated at $6 billion for 2007. In practice, an increasing share of the advertising budgets of major European and Asian businesses is being shifted from traditional media (most of the time local) to online advertising media (mainly American). In the end, an increasing share of consumption expenditures of Europeans and Asians is captured mainly by companies headquartered on the West Coast of the USA.

There is an interesting calculation from Rory Cellan-Jones, a blogger for the BBC[30] based on Google's earnings release in April 2008 reported by Saul Hansell in the New York Times Technology blog: "Google earned $803 million, about £407 million, in Britain in the first quarter. If you assume that rate won't grow, that makes £1.6 billion for the year. And since Google's British earnings are up

40 percent from a year ago, it is a safe bet it will grow. That means Google will overtake the ITV television network[31] as the biggest seller of advertising in Britain this year, Mr. Cellan-Jones figures. ITV sold about £1.5 billion of advertising last year. Britain's biggest commercial television business—the original "license to print money"—is about to be overtaken by an American upstart which only arrived in the UK in 2001." Google sells more advertising than any company in the world. .... Excluding the payments it makes to companies that display its ads, Google's total ad revenue will be $15.7 billion in 2008 [according to Morgan Stanley]." By comparison, "Time Warner, the largest media company in the world, earned $8.8 billion in advertising revenue last year. Viacom had $4.7 billion in ad revenue last year..." In addition, "Google [was] named world's No 1 brand for 2007 and topped a list of the world's 100 most global powerful brands before GE, Microsoft, and Coca Cola. It estimates its value to be $86bn (£43bn).[32]

Make no mistakes; there is nothing unnatural about these financial transfers. They are as legitimate as any of those described in Friedman's The *World is Flat* or even earlier, by the British economist David Ricardo's theory of comparative advantage[33]. If Chinese manufacturers are better than those in other countries, value will flow to them. If Indian companies can manage Information technology better than their counterparts in other countries, value will similarly flow to them. If American online advertising companies can provide superior returns to European or Asian advertisers than their counterparts in those regions, revenues will flow accordingly. American companies are dominating online advertising because they currently have built a comparative advantage based on their ability to invent and innovate as well as to mobilize exceptional talents and skills. Advertising is a peculiar business. Europeans use American online advertising companies because each euro they spend with them is ostensibly returned in the form of securing increased sales to their customers at the lowest cost possible. There is no special mystery here. The success of Japanese cars in the US made US car manufacturers innovate and regain market share, in the US and the world. This situation in online advertising will similarly trigger in other parts of the world development of disruptive innovations in the field. This in turn will enable new rich creative applications and services to be developed. We can already see that in the online business, Chinese firms are beginning to occupy leading market positions in domestic markets as is the case of qq.com, sina.com.cn, and Baidu, the latter competing directly with Google for the search market. Similar trends are visible elsewhere: the social network Bebo is well positioned in the UK, while in Russia Yandex competes with Google and V Kontakte with Facebook. Finally in France, Dailymotion is a tenacious competitor to YouTube.

*

If the digital revolution, the Internet, and mobile technology are the sinews of second life networks and if immersive social networks and abundant content constitute their key uses, then "free" is their business model. There is little possibility that the telecommunications value chain players facing these deep paradigm shifts will remain isolated and static. At a time when the international financial system is fragile and there are signs of a significant slowdown in the global economy, the value chain of telecommunications is showing clear signs of exceptional vitality and activity. Ultimately, this is the most tangible manifestation of networks' entry into their second life.

CHAPTER 5

A Premonitory Turmoil

Several periods in the history of the telecommunications ecosystem have been marked by successive waves of mergers and acquisitions. The last one took place at the end of the 1990's and was associated with the so-called Internet bubble or dot.com bubble. This latest series of consolidations are distinguished by their scope; they affected all sectors of the ecosystem, and coincided with a high level of economic growth in the economies of Europe and the United States.

A new wave of mergers and acquisitions gathered impetus in 2004-2005. By 2007, however, it had taken an unexpected direction. First, even though the global economy saw record mergers and acquisitions across all its sectors in this period, the US "sub-prime crisis"[1] had largely suppressed this trend except in the telecommunications sector. Part of the reason was that the telecommunications sector provided a solid refuge for the worldwide glut in investment liquidity given its stability and distance from the banking sector. This interest in telecommunications was further reinforced by petrodollar flows coming from the Middle East funds, and has led some to believe that there is a new investment "bubble" forming, this time around telecommunications and its allied technologies. While there is no available evidence at this time that definitively supports such as possibility, it is clear that the second life of networks provides an open-ended potential that certainly justifies the positive assessment of the sector by investors.

Secondly, unlike the case of the "Internet bubble," the changes we are seeing in telecommunications are occurring at a time when world economic growth is slowing (at least in developed countries, what growth there comes from burgeoning economies of emerging countries). These two unrelated trends—on one hand a slowing world economy and on the other an effervescent and vital telecommunications value chain—are indications of the fundamental structural[2] change that augurs the second life of networks.

Thirdly, a growing number of the latest mergers and acquisitions sweeping the telecommunications sector are of a new and different nature. In the past, the consolidations were associated with companies wanting to develop and expand their core activities to gain economies of scale and market share. This growth was pursued through external or non-organic means; but remained well within the confines of their position in the value chain. There have been a growing number of unforeseen acquisitions or mergers mixing players from different parts of the value chain.

Horizontal Amalgamations Among Operators and Equipment Manufacturers

Examples of "natural" or "horizontal" consolidations, those occurring within a layer of the value chain, abound, and their numbers have been growing since 1990. The most significant instance of this trend is in the United States telecommunications sector with what was effectively the apparent reconstitution of the Bell System. In 1984, a landmark antitrust decision divested the Bell System. It resulted in one long distance company (AT&T) and seven regional telephone companies (RBOCs, or Regional Bell Operating Companies, known as "Baby Bells"[3]).

In 1994, in order to improve its access to the end-user and diversify into wireless, AT&T purchased the mobile operator McCaw

Cellular Communications for $11.5 billion. Subsequently, AT&T decided to enter into another kind of fixed network, cable[4], through its purchase of two US cable operators, MediaOne (owned by the baby Bell US West) for $62 billion and TCI for $48 billion. With these acquisitions, AT&T became the largest cable operator in the US. The new company created by these assets, AT&T Broadband, was forced a few years later to sell its cable networks. Chiefly because it could not get market acceptance of a bundled television, telephone and Internet offer. AT&T exited the cable business after it spent $106 billion to get into a business that melted down in the Internet bubble burst and left it holding $65 billion in debt. Wall Street wanted a new story, and Michael Armstrong's dream of convergence 1.0 turned into a restructuring story that led it to spinning out wireless and selling its cable unit.[5]

More recently in a turn-around of epic proportions, SBC Communications, one of the RBOC's created by the divestiture of the Bell System[6], purchased AT&T in 2005 for $16 billion, then BellSouth in 2006, another RBOC, for $67 billion complemented the "reconstitution" of a part of the Bell System. SBC's earlier purchase of two other RBOC's, Pacific Telesis in 1997 for $16.5 billion and Ameritech in 1999 for $60 billion.

In the UK, where deregulation started in 1982, the growth phase of the incumbent carrier BT (British Telecom was privatized in 1984 and renamed BT in 1991) manifested itself internationally in the early 1990's. In 1994, BT and the US operator MCI launched a joint venture. Two years later, BT announced its plans to buy MCI, which elicited an ultimately successful counter-offer in 1997 by WorldCom[7].

In continental Europe, following deregulation of the telecommunications sector in 1998, the horizontal consolidations began in 1999. As early as the mid-1990's there were (non-stock) based cross-border alliances among operators to address multinational enterprises. One example was Global One created in 1996 which consisted of France Telecom, Deutsche Telekom, and Sprint. Stock-based horizontal consolidations that did occur near the end of the

1990's were also cross-border due to the regulatory initiatives put in place to promote competition in both consumer and enterprise sectors. These measures favored increasing the number of players rather than their consolidation in each of the countries. For example, France Telecom, the incumbent carrier in France, acquired between 1999 and 2000 operators and licenses valued at close to 100 billion euros in a number of countries outside of France including the acquisition of Orange in the United Kingdom.

Similarly in the equipment manufacturing layer of the value chain, the end of the 1990's saw many horizontal consolidations. The French manufacturer Alcatel purchased DSC for 4 billion euros in 1998 and Newbridge for $7 billion two years later in 2000. In the US, Lucent bought some 30 companies including Ascend Communications for $20 billion in 1999.

This trend continued, albeit with a pause, following the burst of the "Internet bubble" in 2000. In 2005, Verizon purchased the number two long distance operator MCI for $5.3 billion. Near the end of 2004, Sprint purchased Nextel for $35 billion. In Europe this trend was illustrated by the purchase by Spanish carrier Telefonica of O2 for more that 25 billion euros following its acquisition of the Spanish mobile operator Telefonica Moviles (under the Movistar brand) for more than 3 billion euros. Similarly, Deutsche Telekom in Germany reintegrated its mobile subsidiary T-Online during the period between 2005 and 2006 for almost 3 billion euros. The Italian carrier Telecom Italia followed suit in 2006 reintegrating its mobile subsidiary TIM for more than 20 billion euros. In 2005, France Telecom-Orange took an 80% interest in the Spanish mobile operator Amena for more than 6 billion euros. In France, the mobile operator SFR purchased in 2008 the fixed line operator Neuf Cegetel.

In the telecommunications equipment sector similar horizontal consolidations included the acquisition in 2005 of Marconi by Ericsson; the merger in 2006 of Alcatel and Lucent[8] (valued at some 11 billion euros); the 2006 Nokia-Siemens Networks joint-venture which resulted from the merging of their respective network divisions (valued at about 25 billion euros); and finally, the merging in

2005 of Sagem and Snecma, called Safran in France and capitalized at some 6 billion euros.

This horizontal consolidation trend was also present in value chains in adjacent sectors. In software, it includes the purchase of Cognos by IBM for $4.9 billion (one among the 40 acquisitions by IBM since 2000); the acquisition of Business Objects by the German company SAP; and the acquisitions of PeopleSoft ($10.3 billion), Hyperion Solutions Corporation ($3.3 billion) and BEA Systems ($8.5 billion) by Oracle who during the last four years made 40 acquisitions. In the Information Technology hardware sector, HP made 21 acquisitions between 2005 and 2007[9]. In May 2008 the Wall Street Journal reported that HP "said it agreed to pay about $13.9 billion, or $25 a share, in cash to buy Electronic Data Systems Corp setting the stage for a challenge to IBM's dominance in computer consulting and services."[10]

Mergers Between "New" and "Old" Players in Different Layers of the Value chain

We are also seeing a parallel trend in mergers and acquisitions betweens parts of the value chain representing very different skill sets, often between "older" and "newer" players. Equipment manufacturer Cisco, who originally sold only network equipment (routers, bridges, etc.), purchased Scientific-Atlanta, a manufacturer of television set-top boxes, in 2005 for $6.9 billion. This allowed Cisco to enter the residential cable television market[11]. In 2007, with a view to expanding its offer to enterprises, Cisco acquired Webex, supplier of teleconferencing and other collaborative software, for $3.2 billion.

In the equipment sector, the movement from old to new core competencies was also evident. It took a particular form, as hardware present in other value chains (computers, audiovisual and video games) became "communicating" (by direct integration of modems and decoders in existing hardware or even developing new hardware that communicated directly). This has resulted in a growing amalgamation of the equipment layer of the telecommunications value chain with those of other value chains.

In this respect, the case of Apple is canonic. Apple had already begun its expansion out of the computer equipment value chain by allowing its Macintosh computers to communicate[12], and by developing and marketing a web-based music player, the iPod, and most recently a mobile telephone, the iPhone. In April 2008, Apple bought a microprocessor chip designer called PA Semi.

This repositioning of Apple Computers included a re-branding of the company as simply Apple. Meanwhile, Apple's market capitalization soared in 2007 beyond that of IBM, which found itself confined to horizontal acquisitions in the information technology space.

In the service layer meanwhile, Google—whose core competencies are search engines and online advertising—acquired YouTube, an online user generated video site, for $1.65 billion in 2006 as a way to increase and diversify their audience. This followed in 2007 with the acquisition of Postini, a communications security firm which increased Google's enterprise market presence. It also acquired GrandCentral, an Internet telephony platform provider in the same year. Google announced at the end of 2007 an initiative called "Open Social" designed to provide a basis for federating disparate social networks. Many perceived this as a response to the growing hold on this space by Facebook (in which Microsoft is an investor). Google also launched its Friend Connect initiative for developers in 2008.[13] Friend Connect is a new data portability product designed to propagate social connections throughout the Internet. A week before, Myspace launched MySpace Data Availability and Facebook launched its Facebook Connect which are similar initiatives as Friend connect.

On December 20, 2007, the Federal Trade Commission (FTC) authorized the acquisition by Google of online advertising firm DoubleClick for $3.1 billion, which took place in March 2008. Microsoft, AT&T and media giant Time Warner banded together to petition the US government in April 2007 to block this acquisition

Date	Company	Business	Value (if available)
4 January 2007	Xunlei (4% acquisition)	File sharing	$5,000,000
16 February 2007	Adscape	In-game advertising	$23,000,000
16 March 2007	Trendalyzer	Statistical software	
17 April 2007	Tonic Systems	Presentation program	
19 April 2007	Marratech	Videoconferencing	
13 April 2007	DoubleClick	Online advertising	$3,100,000,000
11 May 2007	GreenBorder	Computer security	
1 June 2007	Panoramio	Photo sharing	
3 June 2007	FeedBurner	Web feed	$100,000,000
5 June 2007	PeakStream	Parallel processing	
19 June 2007	Zenter	Presentation program	
2 July 2007	GrandCentral	Voice over Internet Protocol	$45,000,000
20 July 2007	Image America	Aerial photography	
9 July 2007	Postini	Communications security	$625,000,000
27 September 2007	Zingku	Social network service	
9 October 2007	Jaiku	Micro-blogging	

Google acquisitions in 2007

Source: wikipedia.(http://en.wikipedia.org/wiki/list_of_google_acquisitions)

as it would give Google a dominant position in the online advertising space.

Yahoo! was also active and acquired in 2007 BlueLithium, an electronic ad distribution network, Right Media, an online advertising marketplace as well as Zimbra, an enterprise messaging firm, giving it presence in the enterprise market and in particular in its small and medium segments. The online auction firm eBay, for its part, purchased in 2005 Skype the pioneering Internet telephony company for $2.6 billion. Previously, eBay acquiered the online-secured payment site Paypal, in 2002, for $1.5 billion.

These players' moves have allowed them to update their offers while remaining centered in their original positions in their value chains. They could have equally accomplished this through organic (internal) means. Doing so requires the companies' leadership to have the vision to anticipate changes in their markets and the ability to develop internally the means to react to them quickly and appropriately. In this regard, Amazon stands out as a special case. Created in 1995 it is one of the few companies, along with eBay and Yahoo! that survived the burst of the Internet bubble in 2000 and therefore merits closer examination.

Amazon originally was conceived as an online bookseller. Today, Amazon markets all kinds of goods and services in addition to books including music CD's, DVD's, furniture, certain categories of food, toys, video games, home electronics as well as software. Amazon launched in September 2006 a video on demand ("VOD") service developed for PC's and television sets.

Amazon today sells a range of service that go well beyond their original core competencies. These services leverage Amazon's "back-office" software and IT infrastructure, which has consistently differentiated them from other players in this domain. This has allowed it to become a leader in over-the-Internet provision of software development and IT management platforms.

Amazon.com sells on a wholesale basis its e-commerce IT infrastructure and networks. A number of major brands use this service to drive their e-commerce sites and distribution channels.[14] As of 2008, Amazon has also made available on a wholesale basis its electronic payment system "Amazon DevPay."

Since 2002, Amazon has opened access to its IT applications across a well-developed set of open interfaces available on their Amazon Web Services (AWS) developer portal. This collection of web services acts as "anchor points" to applications residing on Amazon's IT platforms, which then can be used as "building bricks."

Developers are able to select, across an Internet interface, these bricks to construct applications that precisely fit their requirements. This represents a new division of labor that allows entrepreneurs to focus on their core competencies and to off-load onto Amazon the majority of their IT tasks. At the start of 2008, there were some 10 different Amazon Web Services[15] available and a registered community of about 290,000 developers using them to varying degrees. Amazon is also contributing to the growth of "Cloud Computing" (see below) which will further reduce the cost barriers entrepreneurs face when they create new companies. Amazon's platforms' allowed the organic development of a number of avant-garde web services that had nothing to do with their original core competency of online bookseller. This has allowed them to evolve their value proposition not only for their traditional clients but also to meet the needs of new segments.

As was posted on ZDNET Larry Dignan asserted that "... In fact, Amazon's real business down the line will be its cloud services. (...) Books will be just a front to sell storage and cloud computing[16]."

Quentin Hardy recounts that Zillow, an Internet site employing 165 people and specializing in US real-estate information and analysis, "...have been anxiously watching housing prices collapse. Hoping to spice up its offerings to a discouraged consumer, Zillow recently recalculated the values on 67 million homes. The database of figures that took up 4 terabytes [equal to 4 thousand billion characters or signs] of memory. The company figured it would need six months and millions of dollars to make it happen. Instead, Zillow ran the job over the Internet, on 500 computer servers rented from Amazon.com. It took only three weeks and cost less than $50,000."[17]

Based on the same principles used by Amazon, eBay also opened up its IT platform as a web service in order to provide the tools to third parties to stimulate complementary commerce[18]. As has been expected, Google finally announced in April 2008 its own long awaited Internet "cloud" development platform for hosting, computing, and storage called Google App Engine.

The Growing Number of Vertical Combinations

While horizontal consolidations were being actively pursued by network operators and equipment manufacturers, and players in all the layers of the value chain moved to the greatest extent possible towards new and innovative value propositions, numerous vertical, "inter-layer" mergers and acquisitions have accompanied the entry of networks into their second lives.

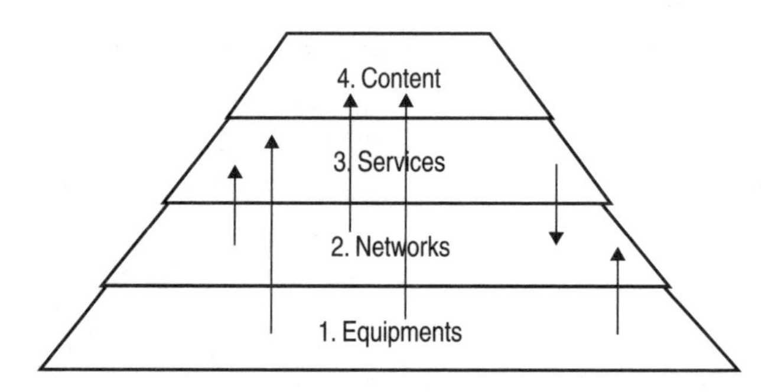

The blending of the value chain's layers

We are witnessing a veritable "blending," a verticalization of the telecommunications value chain. Concretely speaking, it has resulted in a growing number of "silos" constituted of a large number if not all of the layers of the value chain. It involves companies that are initiating acquisitions as well as those that are their targets.

The 1970's already saw some vertical transactions in the telecommunications sector. These were, however, small in number and limited primarily to transactions between telecommunications operators and equipment manufacturers or sometimes between these operators and computer equipment manufacturers (that is to say, "cross-value chain" transactions). At the same time, each of the two rival siblings, telecommunications and computers, were looking to position themselves in each other's markets. This was motivated by the possibility of gaining larger market shares in the enterprise and government institution markets, both major consumers of these two sectors' products and services.

As early as 1975, IBM attempted to enter the satellite telecommunications business through a partnership with Aetna Life and Comsat to launch in 1980 SBS (Satellite Business System) in the US. The goal was to provide satellite transport services for voice, data, and business television to enterprises and government institutions.

In 1991, AT&T entered into the commerce and finance hardware market by acquiring the cash register manufacturer NCR for $7.4 billion. NCR subsequently also began manufacturing computerized commercial hardware such as bank ATM's and Point of Sale (POS) terminals. AT&T remained the primary client of its NCR subsidiary which it temporarily renamed AT&T Global Information Solutions. Shortly after, in 1995, it was spun off because: "...[A]dvantages of vertical integration [which had motivated ATT's earlier acquisition of NCR] are outweighed by its costs and disadvantages...[T]o varying degrees, many of the actual and potential customers of Lucent and NCR are or will be competitors of AT&T's communications services businesses. NCR believes that its efforts to target the communications industry have been hindered by the reluctance of AT&T's communications services competitors to make purchases from an AT&T subsidiary."[19]

Microsoft also joined this early wave of vertical consolidations by obtaining in 1997 an 11.5% share of the cable operator Comcast for $1 billion. The goal was to facilitate adoption of its MSN portal. This was followed up by a $5 billion investment in AT&T, then in 1998, a more modest amount, $200 million stake, in the US operator Qwest[20] as a way to promote Microsoft's enterprise e-commerce services and network products. In 1999, Microsoft participated in mobile networks by taking a $600 million stake in Nextel (US mobile operator subsequently purchased by Sprint to create Sprint-Nextel) as a way to promote its mobile version of the MSN portal. Microsoft also pursued a similar strategy in Europe by investing in cable networks in Portugal with TV Cabo and in the UK with NTL, for a value of $500 million. Despite these investments, Microsoft's success in the network layer of the value chain can best be characterized as limited.

Since the beginning of networks' second life in 2004, the instances of vertical consolidations have multiplied across all layers of the value chain[21].

In the equipment sector, the case of Apple's incursions into the service and network layers of the value chain is probably the most advanced example of vertical consolidation. Apple belonged initially to the computer value chain. It prepared its value chain migration first by developing its own online music content distribution platform iTunes, then the iPhone, launched in 2007. The iPhone, its "home made" smart mobile phone, allowed Apple to participate in networks, at least financially. Apple developed a new business model with network operators, much as it did with the music industry in the case of iTunes. Initially, Apple sold an exclusive, country-wide right to operators to sell the iPhone. The operators bought the equipment which they resold and gave Apple a part of the usage revenue generated by the iPhone.

Another example is Nokia's purchase in 2007 of Navteq, a navigation service company, for $8.1 billion just after having acquired the German mapping and navigation company Gate 5 in 2006, as

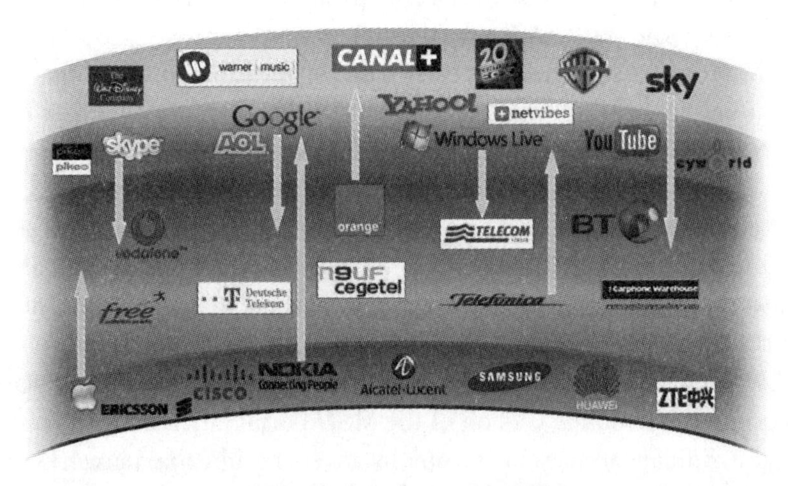

Moves by key players since 2004

well as mobile music service company Loudeye for $60 million, online advertising company Enpocket, geographic location based services company Picto, and picture and video sharing service Twango.

During Nokia's annual shareholder meeting in May 2008 in Helsinki, CEO Olli-Pekka Kallasvuo told investors: "Our goal is to act less like a traditional manufacturer, and more like an internet company. Companies such as Apple, Google and Microsoft are not our traditional competitors, but they are major forces that must be reckoned with.[22]"

Audiovisual equipment manufacturers that had already begun to integrate telecommunications services into their products also tried to participate in the telecommunications value chain. Panasonic, one of the leading manufacturers of plasma televisions in the world, announced in January 2008 an alliance with Google that would allow consumers to watch YouTube videos—which presupposes an improvement in their quality. Samsung also announced, almost at the same time, a partnership with USA Today to offer the newspaper's Internet-site content for viewing on High Definition, high-end televisions. These initiatives presage a new trend in which television manufacturers are positioning themselves to distribute Internet content.

Another interesting example of this trend is the prediction by the *Los Angeles Times* in April 2008 that[23]: "Sony was about to launch an online video service for its PlayStation 3 videogame console. Sony's service would be provided through the online PlayStation Network". The initial version of the service would include movies and television shows flowing from the Internet to the PlayStation 3. Sony is trying to capitalize on its Trojan horse in the living room, the PlayStation 3. The console is already connected to the TV and the Internet, and has sold more than 4 million units in the U.S. and 9 million worldwide, according to Wedbush Morgan Securities in Los Angeles. The console gave Sony the decisive edge in the battle to establish its Blu-ray discs as the standard for high-definition video in the home, trumping the HD DVD format backed by Toshiba Corp., Microsoft Corp. and others. The new service would position Sony to compete with the growing number of Internet-connected devices and services that deliver video to the TV, including AppleTV, Vudu and Microsoft's Xbox 360 console." Microsoft's Xbox Live service has about 10 million subscribers who can consume online streamed and downloaded video including some in high-definition format.

Garmin International, manufacturer of GPS and navigation devices, announced in January 2008 its entry into telecommunications with its launch of "nuvifone," a new navigation device that includes a mobile telephone with a touch-screen, an Internet browser, and a camera that will take a picture and automatically geo-locate and tag its location[24]. Archos, the manufacturer of digital players has also thrown its hat in the ring by announcing in February 2008 its entry into telecommunications by connecting its devices to 3G/3G+ mobile networks[25].

In the equipment layer of the value chain, Microsoft[26] acquired in 2007 Tellme Networks Inc., provider of a voice recognition-based search engine, a mobile directory service, and a service for computer-assisted customer support. Two months later, Microsoft acquired aQuantive Inc., an online advertising and marketing firm, for $6 billion. This last example is indicative of the major transformative shifts dominating the value chain. It was almost impossible ten years ago to imagine that in 2007 Microsoft would buy an advertising firm. "Ten years ago, there were only 'offline' advertising firms," notes John Hawkins, Managing Partner of VC Generation Partners. "Today you have technology companies entering this space. Overall, technology companies, advertising agencies, and networks are consolidating."[27] Microsoft also entered the social networking space in 2007 by purchasing 1.5% of Facebook for $240 million. This was capped in early February 2008 by Microsoft's proposed buyout of Yahoo! for $44.6 billion, potentially the largest proposed transaction of its type in this sector. However it did not materialize and Microsoft stepped out of the deal in April 2008.

Players in the service layer of the value chain are also trying to insinuate themselves into its other layers. Beyond its organic growth into new innovative areas within its sector, Amazon launched in December 2007 a portable reading device weighing 292 grams (capable of reading books, magazines, newspapers and blogs), with a built-in mobile network radio that allows the seamless download of new content: the "Kindle" is sold for $399 plus the cost of content. The original announcement indicated that 90,000 books were available for download from the Kindle site. Newspapers available include *The New York Times, Wall Street Journal, Washington Post, Atlantic Monthly, Time, Fortune, Le Monde, Frankfurter Allgemeine Zeitung,* and *The Irish Times*. There were also 300 blogs available at launch.

In services, Google represents the best example of this trend. To participate in the network and equipment layers of the value chain, Google is pursuing strategies that go beyond the numerous intra-layer acquisitions described. Google played in early 2008 a major role in structuring the auction of the US radio spectrum liberated by the FCC-mandated end in 2009 of analog television. Google was ready to bid $4.6 billion for this spectrum[28]. The result of the auction did not give Google any spectrum. Rather, it was established carriers like Verizon and ATT that captured the majority of the auctioned spectrum. Despite the fact that some believe that Google was more interested in the adoption of open network provisions by the winners than the spectrum itself: a few weeks after the result of the auction, Google announced[29] it had joined an industry consortium to plan and build a new type of wireless data network in the US. Google is investing $500 million in the project. The other companies involved are Sprint Nextel, Intel, Comcast, Time Warner, and Clearwire. Sprint Nextel will be the majority owner with a 51 percent share. In complement to this initiative, Google announced that it was working with some 30 other players coming from the network and equipment layers of the value chain on an open and free operating system and other software (called "Android") that would be financed by advertising revenues, consistent with Google's traditional business model.

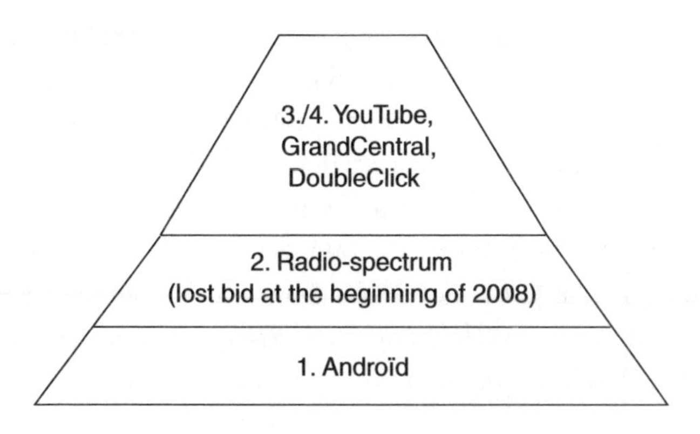

Recent Google initiatives

In the content sector, the purchase in 2005 of MySpace, the social network for youths and adolescents, for $640 million by Rupert Murdoch's News Corp exemplifies the general trend.

In summary, only the network operator layer seems to have so far resisted the trend to verticalization. While certainly, there have been some consolidations in the sector; they have been limited, as we saw earlier, to horizontal consolidations. In Europe, ten years after the deregulation of the telecommunications sector, acquisitions have been limited to new network access service companies in related fields such as Internet television (IPTV) or video-conferencing.

Reasons for the Value chain's Amalgamation

There are five major reasons that explain the amalgamation of the telecommunications value chain.

— The first is that each player in the value chain is trying to get closer to the central assets of any enterprise, the *end-user*. This is because knowledge of users allows a firm to personalize its value proposition and thereby increase the use of its products and services. Furthermore, the resulting visibility the firm gets increases its market presence and "mindshare." This largely explains the growing brand orientation of telecommunications companies today.

Among equipment manufacturers, which represent the layer of players farthest from the end-user, the most representative examples are those of Nokia and Apple. The former launched its own Internet portal, dubbed Ovi, while the latter, with its own stores and music download portal (iTunes) has developed a very strong brand strategy towards its clients, recently reinforced by the launch of its iPhone. In the services segment, Google is reaching end-users through the municipal WiFi network it has operated since 2007 in Mountain View, California. Google also co-invested in 2005 in home-based Broadband over Power Line (BPL) Internet access networks with Goldman Sachs and Hearst Corp.[30] It also invested in FON with Skype and Sequoia Capital. FON turns WiFi capable Internet modem equipment installed in users' homes into relays of a Wi-Fi network open to all FON subscribers.

— The second reason for the amalgamation trend we have seen is economic in nature and has to do with the emergence of the *free business paradigm*. The development of the advertising model is shifting the point of value capture—although not necessarily its point of creation—from the network access to the service layer. Consequently, it has become logical for players in the other layers of the value chain to migrate towards services, especially those with customer assets that can generate advertising revenues. This certainly describes the cases of Nokia, Microsoft, and News Corp.

— The third reason that explains the "convections" or turbulent movements we are seeing in the value chain is that key players in the service layer have aggressively invested in fiber *backbone networks and datacenters* (that is, huge data and telecommunications centers that bring together computers, servers, and storage). Recent developments indicate that this investment trend may soon even extend to mobile networks. Such investments allow these players to manage and ensure the transport of data between their datacenters and the Internet. They accomplish this by creating portions of network infrastructure that parallel the public Internet.

Google also announced in early 2008 that it will be one of the investors in the trans-Pacific undersea cable "Unity" which will connect the US and Japan in order to handle the unprecedented growth in data and Internet traffic flows between Asia and the US[31].

These new Internet trends have made companies like Akamai key players in networks. Not surprisingly, Akamai is a key supplier of Google (including YouTube who previously used Limelight, a supplier of similar networks). Akamai is today building an "overlay" infrastructure over the Internet, which would allow companies like Apple to develop a high performance film rental service that could complement its current iTunes music offering. In this case, this "superstructure" built on the Internet will accelerate the downloading of films (due to statistical technologies that replicate copies of the films in regional datacenters where demand for the film is expected to be the greatest)[32].

— A fourth reason for this mixing of the layers of the value chain lies in the intensification and inevitable *convergence of technologies and services*. One result of this is the acquisition of players with complementary core competencies. This will unify the user's communication environment horizontally by allowing seamless shifting between different networks as well as vertically by allowing, for example, the access of all content and services on all terminals.

The horizontal dimension of convergence is pursued through the consolidation of companies operating different networks. This is illustrated by the numerous acquisitions of ISPs (Internet Service Providers) by what are essentially mobile operators (in France, for example, the acquisition of Neuf Cegetel by SFR early in 2008). The vertical dimension of the convergence described earlier is pushing companies to consolidate with those in layers above and below the ones they currently occupy. Nokia's initiatives cited earlier (the launching of its own Ovi portal as well as its acquisition of photo-sharing and location-based services companies) are good examples of this vertical convergence trend.

— Finally, the fifth reason for the movements we have seen is a more systemic one. It resides in the *mimicking* and hyper-reactivity that always accompanies periods of major strategic realignment of players in a value chain. As soon as one major player moves into a new area (requiring novel core competencies), its competitors will try to follow suit even if it represents a strategy that they had not initially contemplated or ranked as a priority. This mimicking turns into a signal to the market that the strategy emulated may be, in whole or in part, the right one.

Motorola, in order to respond in part to Nokia's acquisition of mobile messaging company Intellisync in 2005,[33] as well as to interdict Cisco's initiatives, rapidly reacted by purchasing the mobile messaging company Good Technology and the modem manufacturer Netopia. This was followed by the acquisition of Symbol Technologies, specializing in barcode scanning and RFID (radio frequency identification) tag technologies, at the end of 2006 for $3.9 billion. In so doing, Motorola has clearly announced its strategy for serving the mobile needs of enterprises through external growth.

Partnerships and R&D 2.0

This rapid overview of reasons explaining the observed amalgamations in the telecommunications value chain as the sector begins its second life would be incomplete if it did not mention the profound metamorphoses that the innovation process itself is undergoing.

The insistent demands of Generation Y[34] for new innovations, be they fundamental or fad, combined with the increase in the number of light and agile players spurred by low entry costs (at least in the higher layers of the value chain), has considerably modified the approach to innovation used by existing players. On one hand, the space for innovation has shrunk. Even if the opportunities for innovation seem to have been expanded with the growth of the Internet as well as of mobile services and devices, the increase in the number of players capable of innovating has intensified the battle to identify innovations and to benefit from them. On the other hand, the cycle of innovations has accelerated in the sense that the time between having an idea and its implementation has shortened significantly due to intensifying competition. In summary, the era of "national," centralized, and long-term innovations is at end, replaced by an innovation style that can be characterized as reactive, even impulsive at times, and global in nature.

In the wake of these changes, the Research and Development (R&D) function in the industries discussed is undergoing a radical transformation. In fact, R&D seems not to be the only or even best thing that rhymes with innovation. Anecdotally, in January 2008, conducting a simple Internet search of books written on innovation results in 7.6 million entries; a similar search on books whose subject is R&D resulted in "only" 545,000 entries.

It is interesting to note that in the 2007 top ten list of companies considered innovative, regardless of sector, by the magazine *Fast Company*; six of the ten—Google, Apple, Facebook, Ideo (a design firm), AliBaba (a Chinese e-commerce company) and Amazon—do not rely on a traditional or "orthodox" model of R&D[35]. In reality, all the sectors that use networks intensely are at the dawn of a new way of doing R&D, one more in harmony with the new "space-time" of innovation, which might be called R&D 2.0.

Several trends have already contributed to the reinvention of R&D. It has been transformed into a set of new activities distinct from all others found in a company. Further, its goals have become more concrete and nearer-term. In a manner of speaking, R&D has become even more responsible for the health of companies and therefore more accountable for their success in an increasingly competitive environment where offer differentiation has become the key success factor. In this context, countries formerly absent from the new technology scene such as India, China, Brazil and those of Eastern Europe represent today a fount of considerable talent, innovation, and development nourishing the rest of the world's innovation ecosystem. This adds even more incentives for enterprises in Western Europe, the United States as well as the technologically advanced Asian countries to evolve their R&D capacities quickly and appropriately. Thomas L. Friedman[36] explains how the "flattening of the world economic playing field" facilitated by networks' growing importance is changing the way trade, financial and human resource allocation, as well as innovation, and R&D are conducted.

What is called R&D 2.0 is best characterized as delivering results comparable to what is available externally, having precisely quantified objectives and delivery horizon for its results, and collaborative (not isolated as was sometimes the case in the past). This last recognizes the truth that it is difficult to innovate alone. This is especially the case in a world where the entrepreneurs seeking to innovate are numerous, less constrained, widely networked, and more agile than ever before. In January 2008, the importance and topical nature of the new collaborative R&D 2.0 was echoed in the 38th Annual Meeting of the World Economic Forum of Davos that had "The Power of Collaborative Innovation" as its theme.

"What happens in R&D, stays in R&D."

Before R&D 2.0

Source: Harvard Business Review

The so-called "cathedral" model of R&D where innovation is pursued end-to-end within the same structure is not an effective model for the second life of networks. It requires the "bazaar" model[37] that is progressively developing in parallel to the traditional "cathedral" approach. A key transformation, especially on the Internet, is the growing predominance of the "bazaar" model that favors aggregation, consolidation, and configuration of building-brick innovations, often coming from outside over hegemonic development from within the enterprise. R&D 2.0 favors decentralization and creativity over organization, process, and planning; vision and leadership over hierarchy. This new concept of R&D 2.0 has distanced itself from the centralized and hierarchic model of R&D 1.0. In doing so, it has become progressively synonymous with innovation and all the freedom, creativity, audacity, and risk-taking that characterize this notion.

There are three levels or "circles of collaboration" that characterize R&D 2.0. A first circle is defined by the enterprise's perimeter: in the "bazaar" model, R&D collaborates more closely with other functions within the enterprise. A second circle is that of the sector or ecosystem broadly defined within which the enterprise finds itself: new services and products are built of a number of innovations some of which are obtained from other players in the value chain including start-ups and universities. The last circle includes all users and independent developers: products are launched as soon as they are usable but not necessarily "finished," in order to obtain input from the product's first users as well as enhancements from developers (using the product's open API's). In this latter practice, we see the concept of "good enough." It applies to the transmission of "packets" of bits on the Internet as well as to the working of the collaborative online encyclopedia Wikipedia. In this case, it permits the testing of products and services by real-world user experience. This allows companies to initiate a process of continuous improvement of their products and services. This new approach represents a wager of sorts by companies on their products and services that is the new "table-stake" of the era of second life networks.

This is a key change enterprises must be willing to make to the early phases of their production processes to combat the ease with which innovations can be copied and rapidly deployed. In this fashion, products and services are adapted dynamically to changing needs of users more rapidly than ever. This is why the current tendency is more one of "buy and deploy" than "build and wait."

There are numerous instances illustrating the insinuation of the R&D 2.0 "bazaar" model in the telecommunications value chain. In the past, major companies like IBM or Motorola departed from their established internal innovation processes in order to develop a product they considered strategically critical to their success. This usually entailed decentralizing their innovation processes by opening them up to the first (intra-enterprise) and second (ecosystem) "circles of collaboration."

In the case of Motorola, its urgent need to re-establish itself in the mobile telephone market, after losing market-share to Nokia in 2003, drove it to put together a small team,[38] located discreetly in a suburb of Chicago, whose mission was to launch development of a new mobile telephone using unorthodox methods. The team consisted of a number of rank-and-file engineers and a recently hired designer. The goal was to develop a device whose major differentiating factor would be its thinness. The team ended up producing an ultra-thin accessory, 10 millimeters thick and 53 mm wide, and in doing so rejected the specifications established by Motorola's R&D which dictated what a mobile telephone should look like (not be wider than 49 millimeters, for example): this was the birth of the "razor" which was marketed under the brand Motorola RAZR. The results were spectacular: several tens of millions of units sold worldwide, which during the period between its launch in late 2004 and mid-2006. This amounted to as many units sold as those of the Apple iPod.

Similarly, IBM, anxious to combat incursions by new personal computer companies, notably Apple Computer, decided in July 1980 to give a small team (called "Project Chess")[39] of about a dozen people, the "Dirty Dozen" as they were called, the task of rapidly creating a personal computer without using the internal design and development processes in place. Notably, these were the very same processes used unsuccessfully in the case of the IBM 5100. The team succeeded in developing the IBM PC in one year by selecting components already available outside of IBM such as the Intel 8080 microprocessor and an Operating System called MS-DOS provided by Microsoft.

While R&D 2.0 is flourishing primarily in the software sector today, it is widely emulated. Cisco "spins-out" internal projects requiring the conditions of a "start-up" environment to succeed and are often subsequently "spun-in" upon reaching their goal. Some companies even acquire products that they deem superior to existing services or to ones developed internally and in various stages of readiness. Google, for example decided to acquire YouTube for its user-generated video service, even though it already had a comparable service, "Google Video." Similarly, Yahoo! acquired Zimbra, a Web 2.0 enterprise email service, even though it had "Yahoo! mail" virtually since its creation.

The software sector's creativity these recent years is largely due to its willingness and technical ability to successfully open its R&D to the third circle of collaboration. This direct implication of the user and developer communities is the core idea behind the open source software movement. More generally, the notion of collaboration that infuses R&D 2.0 is closely associated to the notion of openness: the original need for researchers to collaborate was responsible for the development of the open Internet. Today it is this open Internet together with the open source software that it transports, that is enabling collaboration and stimulating R&D 2.0. Richard Kalgaard, editor-in-chief of Forbes[40], wrote in 2004, "Computing has made the process of innovation, engineering and design implementation in just about every field of science, engineering and business faster, cheaper and more reliable than ever before... This means products can become commodities faster than ever before."

The economist Adam Smith's "invisible hand" mechanism allows the market to reconcile efficiently and optimally the interests of the individual with those of the collective. The twin systems of openness and "free" that permeate the second life of networks themselves act like an invisible hand. In this case, they create an innovation system where the contributions of individuals and developers benefit the wider community of users and developers throughout the world.

In practice, entrepreneurs in the R&D 2.0 tradition are primarily trying to develop solutions addressing their personal needs: reasoning that they may also be of interest to other users. This is why R&D 2.0 can also be called "My R&D." It frees individual creators of services from the constraints and hindrances of the existing technological heritage. This opens the way for ideas to proceed with little friction from conception to design to implementation. Allowing the full participation of the general developer community accelerates the process. The original developer puts down a first brick in a software structure (in the form of a "test version" or "beta version") which is progressively improved and added to as a function of its utilization and user input. "Learning by doing" is one of the key commandments of this new R&D 2.0 tradition.

The majority of Google's applications are apt examples of this trend. Google News is still as of early 2008 a beta version (indicated next to the title on the site's page) as is Google Video. Google Maps and Google Images, however, have both emerged from beta.

In this new universe, a key factor motivating individuals is the fact that each Internet user—each network node, as it were—can be at the heart of a success story featured on the cover of a magazine like *Business Week*, *Time*, *Forbes* or *Fortune*. Entrepreneurs have developed a taste for the personal challenge implicit in the "My R&D" approach. A number of featured company "success stories" brought "cultural hero" status to its founders, some who had just barely graduated from their universities. This is case of Mark Zuckerberg, who founded Facebook in 2004 at the age of 20 while a student at Harvard[41].

Acknowledgment of success during the first life of networks was more discreet: manifested by the number of patents filed and articles published in technical journals. For R&D 1.0, recognition came from peers. In R&D 2.0, it comes from the market. The first life of networks certainly produced "heroes": Bell, Edison, Marconi, and Chappe all acquired significant notoriety in their time. New heroes will also emerge during the second life of networks and even if it is too early to tell if their notoriety will persist into posterity, they will probably be unlike those of networks' first life. They will be drawn from the founders of enterprises such as Google, Facebook, Amazon, Skype, YouTube or Salesforce.com, as well as the myriad of start ups which are not "built to last" but rather "built to flip.[42]" These are the companies that are being acquired by established players in order to rapidly increase their innovation portfolio. As a byproduct, it also interdicts competitors from acquiring companies with key technologies, products. and services.

The telecommunications value chain has undergone a significant change in the way it innovates. The collaborative momentum that characterizes R&D 2.0 has resulted in a multiplying number of links between its players. These are illustrated by a number of agreements to share knowledge and abilities without any formal or explicit financial links in the form of stock-swaps or the like.

The collaboration between companies in a value chain is certainly nothing new. The evolution and spread we are witnessing of the new ways of doing R&D, however, could potentially become the dominant enterprise partnership model of the second life of networks.

Certainly, mergers and acquisitions remain a dominant strategy for growth, allowing the rapid acquisition of experienced employees, market share, and clients, as well as of products, technologies, and expertise. If one is not watchful enough, it might cause problems. During the preceding wave of consolidation of the late 90's, we witnessed examples of cultural incompatibility, management conflicts, and uneven relative company sizes. Moreover, in acquiring companies from other layers of the value chain, certain companies found themselves in direct competition with clients or suppliers, jeopardizing the gains from the acquisition. Nokia and

Motorola both offered online services, and in doing so came into a degree of competition with their primary customers, the mobile operators.

Arguably, the reliance on organic growth as a way to develop new core competencies or expand into new territories is always available. But, it requires overcoming internal barriers. In many cases, competition, and the speed of innovation tend to militate against this strategy as well.

In comparison to traditional alternatives, partnerships offer a number of advantages. As they are more subtle, they force the partners to concentrate on complementarities and synergies as a way to overcome the natural limits faced when expanding to areas beyond what they traditionally know. Their flexibility allows temporary linkages; given that technologies and usages evolve rapidly, today's allies may not be the best ones for tomorrow. Finally, they allow each of the partners to benefit from the mutually created value without having to support the potential costs of restructuration.

*

The exceptional shifts in the telecommunications value chain, are the result of the entry of networks into their second life and of their associated economic evolutions.

These new value chain "tectonics" would not be so important, however, if they were only based on current developments—that is, on things known today—and if in fact, they did not herald a new point of departure, a new discontinuity.

CHAPTER 6

Outlook: Entering the Core of Networks' Second Life

Today we are seeing a unique development in the history of telecommunications. Users are exerting a growing pressure on technology, pushing it to serve new uses. Once, it was technology that tended to push usage, in the sense that upon emergence of a novel technology, users would be challenged to adopt the technology and find uses for it. The invention of the telephone, which ushered in the first life of networks, was not intended to respond to an explicit demand of users—even though we can assume that at some level the idea of communications by voice at a distance was present in people's minds. The telephone quickly became indispensable, and people found many uses for it, which for the most part had not been originally contemplated.

The entry into networks' second life can be interpreted in light of this usual succession of events: technology precedes usages. The immersion of users in "always-on" networks and the new uses that have emerged, such as social networks, are indeed derived from the three technology Big Bangs of the 1970s: the advent of digital technology, the creation of the Internet, and the emergence of mobile telephony. The adoption and growing mastery of these technologies by users, fueled by network convergence, simplification, and even product design, has resulted in the sublimation of physical networks in favor of the human and social ones they enable. This has allowed users to operate in the usage-space rather than in the technology-space, freely creating almost any combination of service and content they can imagine for the future.

Barely a few years into networks' second life, social networks have already become an important locus for users. Many users have become digital content producers, sharing their texts, messages, pictures, or videos, sometimes privately within their interest groups, sometimes publicly with virtually anyone on the Internet. Further, new forms of usage, such as high-definition television and "3D" television and video viewing, which have been the subject of endless promises and false starts during recent decades, are now expected and even demanded by many users as the next "evolutionary step" of networks.

Further development of this parallel digital world, both in space (access available to the largest number of people rather than primarily limited to members of the "Net Generation") and in time (always on connected users), will require new technological breakthroughs. As recently as the late 1990s, the copper networks that permeated first-life networks, the old copper "twisted pairs," were still considered adequate (due to their use by ADSL technology) to provide speeds superior to dial-up service. But today they are no longer sufficient in light of expected information flows. Similarly, the currently available radio frequencies and associated technologies are insufficient to meet the needs of the mobile Internet. Here again, consumption is outpacing and anticipating technologies and thereby accelerating their development. The technological renewal of physical networks will enable all users to seamlessly and continuously access content, services, and social networks, and allow all those who wish to get to the heart of networks' second life to do so.

The Exponential Development of Connectivity

Two main trends can be identified as underlying the current and anticipated increases in network traffic: first, the increase in "smart objects" connected to networks; second, the growing number of video and television services that are responsible for transmitting significant amounts of content over both fixed and mobile telecommunications networks. Both trends are an integral part of the second life of networks because they are the basis for the construction of a parallel digital world, manifested primarily in human and social networks. Communicating objects allow the users who possess them to connect to each other, and the objects also connect directly to other objects, sometimes simultaneously, sometimes independently. Together with the growing video traffic, they demonstrate the growing sophistication and richness of the Internet and mobile networks associated with the professionally generated and user-generated content available today.

The number of actually or potentially communicating objects continues to grow around the world, driven by cost reductions. More than a billion PCs are expected to be active by the end of 2008,[1] the bulk of which are already connected to networks.

The Mobile Internet

At the end of 2007, there were already approximately 3.3 billion mobile subscriptions, according to the company Informa, and 1.3 billion fixed telephone lines in the world.[2] Even if this level of mobile users already seems high, there is still room for considerable growth. With the proliferation of mobile handsets, comes an increase in their connectivity, in terms of speed as well as range of content and services accessed. Further, the fact that mobile phones are increasingly connected to the Internet has already had, and will continue to have, a major impact on the levels of traffic on both networks. As early as mid-2007 in Japan, the number of users accessing the Internet from mobile phones reached the level of those accessing the Internet from the PC *via* a fixed-line Internet connection.

The development of mobile Internet will continue to leverage several forms of convergence, resulting in an increased usability and improved ergonomics. On the one hand, terminals will be required to roam seamlessly from Wi-Fi network access points to mobile cellular networks (and *vice versa*) during an Internet session. This will ensure continuous optimization of the user experience by choosing, at every point in time and space, the best available network without the user's involvement and even without his or her knowledge. On the other hand, simple navigation by moving one's fingers on the surface of a comfortably sized display (the "touchscreen" or simply "touch") will gradually complement and sometimes replace the mini-keyboard and its often tedious[3] and physically awkward succession of clicks required to access desired services. For example, during the summer of 2007, after the U.S. launch of the Apple iPhone, Google and Yahoo! reported a "dramatic" increase in traffic on their websites directly accessed from the home screen of the iPhone. The information accessed included website pages for maps, weather forecasts, stock market prices, and searches (Google reported that its search engine gets 50 times more queries from an iPhone than from any other mobile phone.)[4] The market share of the Internet browser used in the iPhone was estimated at 0.13% in January 2008, representing an average monthly growth of over 20% since its July 2007 launch. Put in other terms, the iPhone actually tripled the volume of data traffic on AT&T mobile phones in cities such as San Francisco and New York during the second half of 2007.[5]

Geolocalization

Another important development came with the imposition of E911 requirements by the U.S. government, which dramatically increased the rate at which mobile phones were fitted with geolocation capabilities (either GPS [Global Positioning System] satellite receivers or phone cell identification and triangulation). According to the firm iSuppli, the number of mobile terminals ready for GPS-based geo-navigation services would reach about 162 million by the end of 2007, seven times the number of car navigation devices.[6] According to David Gill, an analyst with the firm Nielsen Mobile, during spring 2007 mobile users in the U.S. spent on average twice as much for services related to geo-navigation as for music downloads.[7] The growing popularity of maps and their rapid navigation will enable new forms of interaction with online services. For example, clicking on a location will invoke access to associated information. This so-called "Geoweb" will have a much more intuitive—and therefore more easily mastered—interface to drive its powerful underlying discovery and search engine. Beyond this "active" geo-navigation to help mobile phone users to find their way, as it were, increasingly these phones will also be permanently geo-localized. This will allow users, assuming they wish to opt in, to receive "hyper-relevant" services, points of interest and advertising messages pertinent to their real-time geographical location.

The New Communicating Devices

In addition to PCs and mobile phones, objects hitherto lacking two-way communications capabilities are increasingly supporting such capabilities. This is particularly true of the 1.5 billion televisions in service worldwide, which, while lagging behind PCs, are beginning to connect to two-way networks. Previously, televisions were connected uni-directionally, in receive mode, to broadcast TV networks (cable or over-the-air), allowing consultation—such as teletext—and channel selection but no true interactivity. Other electronic devices are also increasingly connected—for example, game consoles (which already allow high-definition video downloads). According to the company Comscore, the number of online gamers worldwide (on games consoles and PCs), which reached 217 million at the end of 2007, is expected to grow at an annual rate of about 20%. It is probable that in the near future, with all or most objects in the home able to wirelessly communicate with each other, the "digital home" complement to second-life networks will finally become a reality. In this true "digital home," appliances of all kinds will be remotely controlled, pictures in digital frames will be remotely changed, and home state-change signals will operate throughout.

Sensors and RFID: Machine-to-Machine Communications

Finally, the deployment of sensors (readers) and electronic radio tags such as those found in RFID technology [Radio-Frequency Identification], will connect our entire environment, creating a kind of "global sensorium."[8] These micro-electronic circuits can be embedded in mobile phones, but also in any other device. They can continuously transmit information about the environment, health conditions of living beings, the state of equipment or merchandise, as well as identity, all without requiring direct physical contact with other devices. They will trigger the development of new applications linked to the context of users, since they can continuously measure multiple parameters. The emergence of these sensors, RFID and other, is expected to generate a wealth of information that networks will have to deliver reliably because of its often critical importance. For example, in the health sector, sensors can be used for medical surveillance or to transmit biometric data directly from patients; in such uses, information reliability is clearly critical and requires service-level assurance.

There will also be many applications in the field of m-commerce (mobile commerce). For example, barcodes of a new type are already appearing in magazine product ads. Such codes can be "recognized" by a magazine reader's mobile phone; *via* the mobile phone, the user then has direct access to related online commerce sites, where he or she can conduct in a few seconds further interactions or transactions with the product and brand.

Most notably, this expanding cloud of sensors and radio tags will mark the advent of communications between machines, known as M2M (machine-to-machine). Forecasts even suggest that there may soon be more machines communicating among themselves by radio than there are mobile subscribers. M2M traffic is expected to take off, starting in 2009 according to analysts, with a volume of up to 4 trillion connection requests by 2010.[9]

An RFID chip

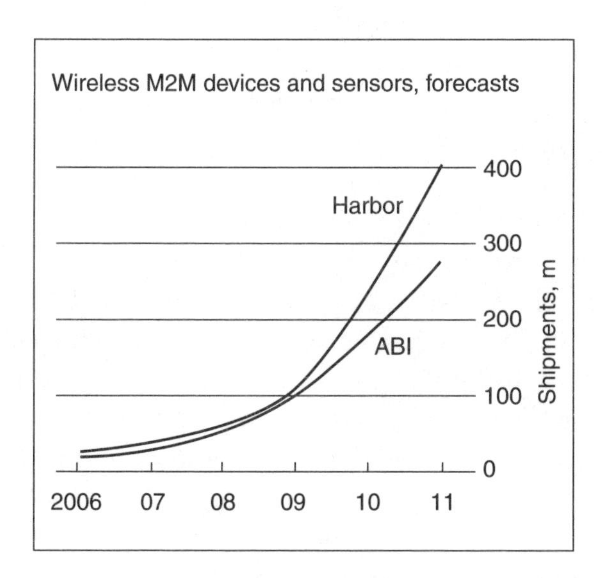

On the right wavelength

Source: Harbor Research, ABI Research in *The Economist*, April 26, 2007.

In its report "The Internet of Things," the International Telecommunication Union (ITU) describes a new era of ubiquitous computing and communications: "A new dimension has been added to the world of information and communication technologies (ICTs): from anytime, any place connectivity for anyone, we will now have connectivity for anything." According to the report, the objects may even make decisions without interrogating the network, as for instance when a simple response to external stimuli is required. This is precisely what is meant by "smart" objects.[10]

One implication of the proliferation of communicating objects is that the IPv4 Internet addressing protocol currently in use will exhaust its "mere" 4.3 billion addresses sometime between 2010 and 2017. IPv4 is being gradually replaced by IPv6, which has a much larger number of IP addresses—about 10 raised to the power of 38![11]

The Growth in Video and TV

The second major trend that explains the increase in traffic on networks has to do with the evolution of the content carried. The always-on Internet user experience that accompanied broadband access encouraged increased consumption of all types of rich content. It is akin to the enhanced TV experience, with the added dimension of the Internet-centric experience of traditional web searching and browsing. Users are consuming more—and creating more as well—since they are always on, connected at faster rates than they were during the first life of networks. In addition, more and more video content and TV programs are available on the Internet, allowing space and time shifting and raising expectations regarding content sophistication.

Internet video for the PC has grown rapidly, with sites like YouTube, Joost, or Dailymotion allowing anyone in the world to upload any type of short video they created or copied from other sources (sometimes illegally). A study by Pew/Internet in early 2008 showed that 48% of Internet users in the U.S. in 2007 visited at least one video sharing website. This is up from 33% a year earlier, representing an annual growth rate of 45% (this growth rate is 59% for women and 58% for the 50–64 age group). In terms of traffic, the numbers are even more impressive. During a typical day, the traffic to these sites nearly doubled in one year.[12] Another study by Pew/Internet, this one conducted in 2007, showed that the consumption of video on the Internet is subject to true "viral propagation," as 57% of those who consume video on the Internet send links to these videos to others. In this context, young people aged 18 to 29 appear to be the most "contagious" vector, with 67% of those who consume videos on the Internet sharing links to them. According to the same study, 8% of Internet users already upload their own videos on video posting and sharing sites.[13]

Video sharing, like music file sharing in the early 2000s, is occurring mainly on peer-to-peer networks, not all of them operating legally. According to Cisco, in 2007 such networks accounted for 50% of non-private general Internet traffic worldwide, the largest single share.

The consumption of video over networks is expected to accelerate in the coming years. Although in 2006 legal downloads of video files in the U.S. accounted for only 1% of the $24.5 billion DVD sales and rental market, this share is expected to reach 10% in less than a decade.[14] However, on YouTube, for example, videos are short (a few minutes) and relatively low quality. In September 2007 average usage was 15–20 minutes per day per user (which still represents a distribution of 100 million video streams per day). In March 2008, Philip Inghelbrecht, strategic partner development manager, at YouTube Inc, said that "Ten hours of fresh content is uploaded to YouTube every minute."[15]

Similarly, short video clips represent 70% of the volume of viewed video on mainstream media sites, such as CBS.[16]

Commercial video-on-demand (VoD) for the PC has been launched by Apple with AppleTV and Amazon with its Unbox. These services allow users to watch commercial videos of any length in a relatively comfortable fashion. Some special Internet TV boxes, such as the Slingbox, permit viewing of live or recorded TV programs in another room or another part of the world on a PC *via* the Internet.

What is expected to really give a boost to viewing long-form TV programs on the Internet in the coming years is the improved user experience afforded by Internet television (IPTV). According to John C. Dvorak: "In the ideal IPTV scenario, anyone at any time can watch ... [any television program] ... in the entire world that is deliverable over the Internet ... IPTV will be able to deliver all the content in the world, on demand, once it is universalized. ... The key to IPTV is that ... the number of shows will be enormous and availability will not be dictated by time of day or geography."[17]

The firm Multimedia Research Group projects that the number of subscribers to IPTV in the world will rise from 13 million in 2007 to 73 million in 2011. This represents an average growth of 40% per year. Europe is leading this trend, even if Asia will catch up with Europe in the not so distant future.

Improvement of the transmission quality associated with television flows will be one of the major factors driving traffic increase on second-life networks. In particular, television on the Internet will be in a position to evolve to high-definition (HD), Super HD (2160 vertical lines), Ultra HD (4320 vertical lines) television broadcasts[18] which could require as much as 256-480 Mbps of bandwidth, then 3D-like, and finally, in time, to "real" 3D. Real 3D will be quite different from current rudimentary attempts; for example, special glasses will not be required since spatial depth will be perceived regardless of viewing angle. 3D is probably what most captures users' imaginations because it introduces a new dimension to the video experience, one that goes far beyond improvements that have already been experienced in image definition or color rendering. Those who have already had a chance to see a video in 3D understand the magnitude of this coming change in the home digital universe.

3D Television...

Source: Serguei Tiounine, *Kommersant*, Moscow
(reprinted in *Le Courrier International*, 8 November 2007)

Video will also invade mobile networks. Consumption and transmission of videos as well as personal mobile television are expected to grow rapidly because of the increase in data bandwidth on mobile networks and the improved user experience of mobile handsets with larger and better quality screens.

In addition to video, other, less bandwidth-hungry uses will continue to grow, including Internet telephony, simple web page browsing, and other usage behaviors that are already quite common. For example, the "memory" of users is increasingly shifting from their PC's hard drive to the Internet. It is not uncommon for users to search on the Internet for documents or content they know is on their PC but that they simply cannot find as quickly or as efficiently there as on the Internet, given the perceived limitations of PC–resident search algorithms and indexing schemes.

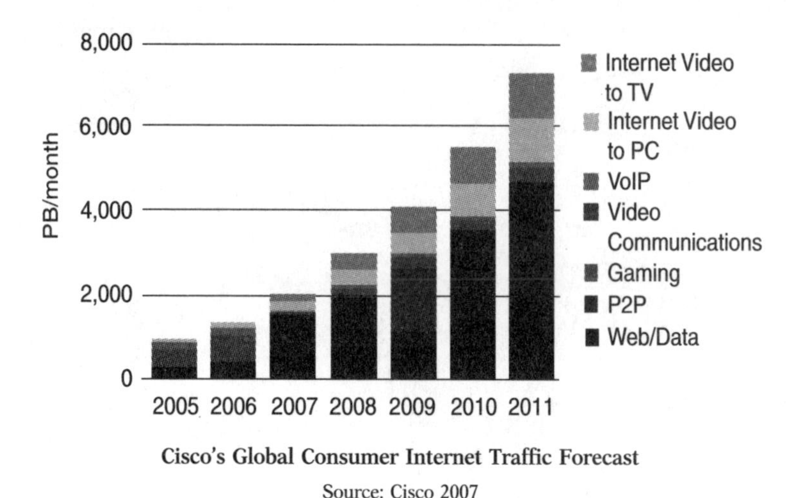

Cisco's Global Consumer Internet Traffic Forecast

Source: Cisco 2007

Towards the Limits of Networks' Capacity

Even the most conservative projections predict that as networks enter their second life there will be a concomitant increase in their use. This will result from the growing number of users who are taking the leap into human networks, into a digital parallel world where users are permanently in communication (both actual and potential) with each other. In this parallel world, the "communication bond" among users is constantly being enriched by the generation of increasingly sophisticated and multidimensional content, such as video. And this is happening independently of geographic locations.

The complex of underlying physical networks has become transparent to users, a development that has enormously improved user experience by "increasing 'availability' and decreasing 'visibility' of processing power."[19] All forecasters implicitly assume that these substrate networks will handle the considerable increase of traffic generated by the explosion of new usages we are currently witnessing. But if the second life of networks has indeed been built on the physical skeleton of first-life networks, consisting mainly of copper-ADSL and 2.75 Generation mobile networks (using Edge technology, for example), that structure definitely will not have sufficient capacity to handle widespread second-life networks. In short, the bright future of the second life of networks just described, is by no means guaranteed.

Since it has become possible for anyone to transmit information of any kind on networks, information traffic has grown at a rate consistent with the known exponential growth characteristics of information technologies.[20] With increased computing power, data storage, transmission, and screen display resolution, content has developed both in quantity and sophistication, from text to pictures and video. This inevitably means an increase in the volume of digital information bits and bytes transmitted (1 byte = 8 bits, or one character or symbol).

This is why, from the earliest days of the Internet, we have seen Internet traffic, starting from a very small quantity, almost double every year.

Month	Year	Terabytes per month
NA	1991	2
NA	1992	4.4
December	1993	8.3
December	1994	16.3
December	1995	NA
December	1996	1,500
December	1997	2,500–4,000
December	1998	5,000–8,000
December	1999	10,000–16,000
December	2000	20,000–35,000
January	2001	23,000
January	2002	55,000

Worldwide Internet traffic from 1991 to 2002

Source: http://www.dtc.umn.edu/~odlyzko/doc/itcom.internet.growth.pdf

(1 terabyte = 1 000 gigabytes)

Today, the entry into the second life of networks is once again driving an exponential increase in traffic, but this time starting from already high levels. Cisco announced in 2007 that within the next 5 years the total Internet traffic would quintuple, reaching nearly 30 exabytes per month in 2011 (1 exabyte = 1 billion billion bytes, or 1 billion gigabytes).[21] For example, according to Cisco, YouTube, the popular video sharing Internet site founded in 2005, was already in 2006 responsible for the generation of 27 petabytes (equal to 1,000 terabytes) of traffic per month, roughly equivalent to the entire Internet traffic at the beginning of the millennium!

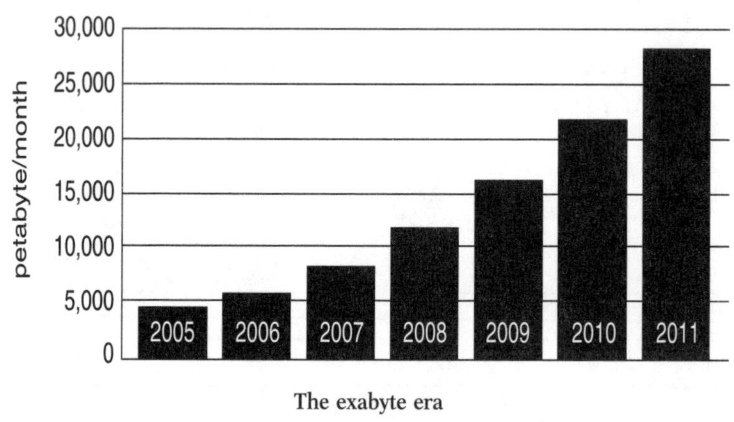

The exabyte era

Source: Cisco Systems Inc.

Bret Swanson in a *Wall Street Journal* article highlighted what he called the "coming of the exaflood"—a huge wave of new content on the Internet that will push the boundaries of the known exabyte world.[22] A significant part of the information on the Internet is buried in what is called the "deep Web", whose size could be 50 times that of the visible or "surface Web": in 2006, it was estimated that there were 900 billion "invisible" pages, compared to "only" 20 billion pages known by search engines.[23] These invisible pages often include contextual content, and databases that are not detected by search engines because they only appear when a specific request is generated by a user. These "created" pages are ephemeral: for example, the pages resulting from the search for a desired airline flight, based on all possible combinations of the points and dates of departure and arrival, rates, and service class, are buried in the deep Web. In April 2008 Google announced it was starting to index the invisible Web often hidden behind web forms.[24]

If we broaden the spectrum to encompass all existing digital information—that is, everything present not only on the visible and invisible networks but also on outside networks, such as CD and DVD—the numbers are even more staggering. A 2007 study carried out by IDC for EMC2,[25] "The Expanding Digital Universe," found that "in 2006, the amount of digital information created or copied in the world stood at 161 exabytes, which is equivalent to 3 million times the information in all the books ever written." The study goes on to say that "between 2006 and 2010, the amount of information added annually to the digital universe will grow by a factor of 6, from 161 to 988 exabytes."

Nate Anderson at Ars Technica indicated that: "Microsoft alone uses an estimated 5,500 terabytes each month just to send movies and television shows to Xbox 360 consoles [and] according to Cisco estimates, MySpace alone moves 4,148 terabytes each month. Even gaming requires big bandwidth. *World of WarCraft*, by itself, moves over 2,500 terabytes of data a month, more than Yahoo!'s US operations. Keep in mind that the entire US Internet backbone transferred only 6,000 terabytes a month back in 1998. Now, a decade later, individual US sites like YouTube require nearly twice that amount.[26]"

Based on another source, "by 2010, the average household will be using 1.1 terabyte (roughly equal to 1,000 copies of the Encyclopedia Britannica) a month, according to an estimate by the Internet Innovation Alliance in Washington, D.C. At that level, it says, 20 homes would generate more traffic than the entire Internet did in 1995. [27]"

As Bruce Mehlman and Larry Irving[28] co-chairmen of the Internet Innovation Alliance explain in "Bring On The Exaflood! Broadband Needs a Boost", an article in the *Washington Post*: "The impending exaflood of data is cause for excitement. It took two centuries to fill the shelves of the Library of Congress with more than 57 million manuscripts, 29 million books and periodicals, 12 million photographs, and more. Now, the world generates an equivalent amount of digital information nearly 100 times each day. The explosion of digital information and proliferation of applications promises great things for our economy and our nation, as long as we are prepared.[29]"

The exponential growth of digital information around the planet, linked to the explosion of usage associated with second-life networks, is primarily driven by the availability of significant technological capabilities. The IT corollary of Parkinson's Law ("work expands to fill the time available for it")[30] postulates that the amount of digital information expands to fill available storage space—i.e., the hard drives in PCs or portable music players, as well as in video players like the Apple iPod or the Archos, each of which contains several tens of gigabytes (a gigabyte is a billion bytes, that is a billion alphanumeric characters or signs).

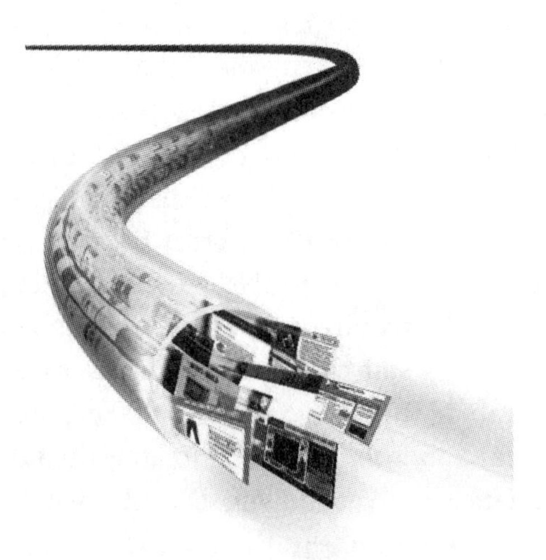

Towards the Limits of Networks' Capacity

Source: "The Exabyte Era," Cisco, January 2008, p. 1. (http://www.cisco.com/en/US/solutions/collateral/ns341/ns525/ns537/net_implementation_white_paper0900aecd806a81a7.pdf)

In addition, the ability to transmit this digital information instantly from one place to another at any time has had a considerable impact on the growth of digital content, such as the immediate mass adoption of email by businesses and individuals. Parkinson's Law can also be applied here in the sense that information and content transmitted over networks are expanding to fill all the available capacity (recall the old saw that "Nature abhors a vacuum"!). This law is akin to the one about markets formulated by the French economist Jean-Baptiste Say (1767–1832): "Supply creates its own demand"[31] (or "the demand for a resource always increases to match the supply of the resource").

What we are seeing today is much different, however. The well-known economist John Maynard Keynes maintained, contrary to Say's Law, that demand creates supply. Today in telecommunications, it seems that Keynes is a better guide than Say. The projected medium-term network traffic levels go well beyond what could be explained by a simple adjustment of usage to available capacity[32]. It seems that usage (demand) is stimulating technological development (supply) in networks, but not in an incremental or linear fashion.

The Imperative of Very High Speed Fixed and Mobile Networks

After the financial bubble burst of the late 1990's, the Internet was seen as oversized for the quantity of content available for transport and exchange. The development of digital photography, online music, and now online video challenges that notion. The most critical element of Internet capacity is the speed of transmission in the access network—the "last mile," the part of the network closest to the user. If certain content did not have to be delivered integrally to users, current network capacity would be more than sufficient for the medium term. The tens of thousand of monthly exabytes of traffic could be delivered using the idle time and transmitting continuously. The real constraint is to be found elsewhere: the dominant uses of second-life networks, such as video on demand streaming, IPTV, and HD television, require instantaneous (real-time) transmission speeds that are difficult to achieve on existing access networks. This is an example of so-called verticalized demand, where the capacity to meet this demand must be available continuously and instantaneously. Horizontal demand allows carriers to spread the capacity over time, thus requiring a smaller capacity to be available continuously. In short, the difference between verticalized and horizontal demand is the difference between transmitting 1 gigabyte of content in a second and transmitting it over a 10-minute slot of time; between receiving a video frame integrally and downloading software, where totality of content is only required at the time the program is installed or invoked.

High-definition television, for example, requires four to five times the network capacity of traditional video (and many times that when compared to the low-quality videos found on user content sharing sites). The emergence of other service technologies will also contribute to the growing demand for more capacity. For example, in January 2008 the government of Japan announced that it was studying a system of "ultra high-definition" television for 2015, using a technology called "Super Hi-Vision" with 16 times finer resolution than HD currently available from the network of Japanese broadcaster NHK.

The increase in bandwidth also allows for a significant improvement in the ease of use of a wide range of other services. For example, it allows faster display of Web pages, which simplifies Internet browsing, and improves the instant messaging service found on most social networking sites. Increased capacity is a crucial development for the second life of networks, so that communications and content exchanges among users can occur in the most natural way, seamlessly, and with virtually no response delay. Such exchanges should increasingly be as fast as face-to-face communications.

In one of his blog posts, Jonathan Schwartz, president of Sun, said that in 2007, it was faster to send a petabyte (1 petabyte = 8 billion megabits) of data from San Francisco to Hong Kong by boat than *via* the Internet.[33] He was alluding to the famous adage from the American writer and scientist Stuart Andrew Tanenbaum extolling the major virtue of magnetic tape as a medium for mass information—that it can be transported faster by mechanical means than *via* the Internet. Tanenbaum said: "Never underestimate the bandwidth of a station wagon full of tapes hurtling down the highway."[34]

More specifically, Jonathan Schwartz calculates that with an Internet access speed of 500 kilobits per second (0.5 megabit, or about the rate of entry-level "broadband"), it would take 507 years to move a petabyte of data! Even with a capacity of 1 megabit per second, it would take 253 years, and a little over 2 years with a rate of 100 megabits per second!

The speed of connections available in the heart, or core, of the network has evolved rapidly since the mid-1990s, when it was somewhere between 45 and 144 megabits per second. Today it is of the order of 10 gigabits per second and expected to increase to between 40 and 100 gigabits per second in the very near future. These bandwidth capacities have been achieved through progress in the field of fiber optics, which makes up the "circulatory system" of all networks.

However, the question of increasing network-access bandwidth has been debated since the beginning of the first nonscientific applications of the Internet in 1993–1994. This is the locus of the network bottleneck: data flows are weaker, and significant investment and operational costs are required to increase transmission rates, since there are at least as many, if not more, access lines as there are users.

At the beginning of the commercial Internet, in Europe, access to the network was primarily provided by telephone companies and made possible through modems (adapters to connect the PC to subscriber telephone lines) and the telephone network. That network operated at speeds between 14,400 bits per second (approximately 1500 characters per second, which was five times the speed of the fax of the day, or 150 times the speed of telex) and 56,000 bits per second. This latter speed was associated with what was known as the Integrated Services Digital Network (ISDN). This network, initially built by European telephone companies, provided data speeds ranging from 64,000 bits per second to 128,000 bits per second. Under certain conditions of use, specific to individual companies, it could reach 2 megabits per second through a process known as multiplexing that required special hardware.

The transmission speed achieved by ISDN was surpassed by so-called broadband technologies. In Europe, the dominant technology was ADSL (Asymmetric Digital Subscriber Line). In Germany, however, Deutsche Telekom used initially its old videotex BTX network, which had ISDN gateways for accessing the Internet. This was similar to the concept of AOL in the U.S.: a closed, online proprietary environment providing access to the Internet. However, this did not eliminate the need for a telephone line and modem by its subscribers. According to the U.S. Federal Communications Commission (FCC), as late as March 2008 broadband Internet access was defined as being faster than the telephone network, providing a symmetrical (emitting and receiving) speed of at least 200 kilobits per second. In practice, the downstream (receiving) throughput of ADSL can reach a few megabits per second, while the upstream (emitting) data rate is usually limited to 1 mega-bit per second. In March 2008, a new definition was introduced that is more consistent with the rapid growth of bandwidth in the U.S.: the required minimum speed to qualify as broadband is 768 kilobits per second, which in practice can grow to 1.5 megabits per second.

Today, at the beginning of the second life of networks, the challenge is to ensure that the opportunities offered are open to all, and to avoid replicating a geographic or generational "digital divide" that limits availability to certain countries or to the "Net Generation." To meet this challenge, fixed-line access networks must increase their capacity by replacing their old copper pairs—which have somehow survived the Internet revolution—with optical fiber to the home or neighborhood. More specifically, there are several types of fiber networks: (1) FTTC ("Fiber To The Curb," i.e., to the users' street), prevalent in the United Kingdom, for example; and (2) FTTH ("Fiber To The Home," i.e., to the users' home), provided in Japan and promoted in France, for example. There is a third, less common approach used in the U.S., called FTTP or "Fiber To The Premises," which means reaching a point somewhere beyond the "curb."

Not only can fiber optics significantly increase current bandwidth speeds—up to 100 megabits per second in most cases—it also provides for symmetrical uplink and downlink speeds to and from users. This symmetrical bandwidth is a central technological feature of the second life of networks: that the user is now a network node, not merely a passive recipient of information.

With the development of fiber networks operating at 100 megabits per second—the culmination of 30 years of continued progress—we again find evidence of an exponential growth law, this time concerning fixed access. Recalling Moore's law concerning the power of microprocessors, Reed and Metcalfe's law of the value of networks, or Cooper's law of radio transmission spectrum efficiencies, this new law, or conjecture, rests on the observation that the maximum bandwidth available on fixedline-access networks in the industrialized countries is generally multiplied by 50 every 10 years. If this observation holds true, it suggests a bright future following widespread fiber deployment.

In the area of mobile networks, emerging second-life usages will require significant improvements in capacity. On the one hand, 3G (third-generation) mobile networks today (called UMTS, or "Universal Mobile Telecommunications System," in Europe) and even those referred to as "3G+" (which includes those using the HSDPA protocol, also called "3.5G") continue to be deployed and used increasingly by companies. On the other hand, the need for new generations of mobile networks is already anticipated, as a result of the growth of mobile Internet, video, and television, as well as mobile social networks. Current 3G networks allow an effective speed of a few hundred kilo-bits per second, and 3G+ networks provide an effective transmission speed of a few mega-bits per second, but even this falls short of what ADSL and fiber permit over fixed-line networks. The new generation of "4G" technology, which in 2008 is still being defined, should allow data rates of several tens of megabits per second. The 4G technology includes development of WiMAX and HSDPA/HSUPA technology within the framework of the Long Term Evolution (LTE) initiative, which brings together many operators and equipment manufacturers.

In the context of wireless, one additional hurdle needs to be cleared in order to upgrade transmission capacity. This is the rationalization of the scarce radio frequencies spectrum. The termination of analog terrestrial TV (in 2011 in most European countries and in 2009 in the U.S.) and the availability of the radio spectrum thus released, together with the development of new generations of mobile phones, will make possible a host of new services, including "personal mobile television."

Another property of networks that will be improved with increased bandwidth is "latency," that is, the required response time for a request to make a round trip through the network. Increased bandwidth can be used to transmit a very large volume of data across the world at a very rapid pace. But some elements of the network may slightly delay the transfer of information (even if it continues to be delivered globally at the same pace).

Applications found on networks today are becoming increasingly rich and complex, requiring near-instantaneous interaction with users; this is particularly true of video games and "tele-presence," which have been the most affected by the uneven transmission capabilities of the network's elements. The quality of future networks will depend in part on eliminating obstacles to uniform, rapid transmission, particularly in mobile networks, which have traditionally been less rapid and responsive in transporting information packets.

Datacenters and "Cloud Computing"

Along with the indispensable implementation of fiber-to-the-home (or as close as possible to the user), other developments will improve the speed of access to relevant information and content in the second life of networks. The new challenge posed by the flood of digital information is leading key players like Google, Yahoo, Microsoft, Amazon, eBay, AOL and Ask to develop computer network architectures capable of handling astronomical amounts of information in parallel to and interconnected with the public Internet. These computers are grouped together in "datacenters," that is, major digital processing and computing centers that represent what we might call the "new factories" of the twenty-first century. These parallel infrastructures allow the large-scale, geographically distributed replication of information. With information infrastructures organized in this fashion, response time to users' requests is made fluid as it travels over the Internet. For example, the "caching" technique invented in the early days of the Internet for web site pages has been repurposed for video management, including that of high-definition video. Thus, content (including videos) that is statistically likely to be frequently requested is replicated from a "central" repository to a local one that is "closer" (in a network-geography sense) to the source of the request. This improves the quality of a video stream. Similarly, a large file download can be achieved more quickly and with less error with this technique. Akamai, a company that specializes in this field, announced that in the U.S., with its own network of "caches" in 750 cities, it can guarantee a flow rate of 100 terabits per second for high-definition video.[34]

The creation of datacenter architectures interconnected to the core of networks may seem to go against the current trend of moving digital intelligence to the edges of networks, that is, to users (especially with user-generated content). In part, this form of advanced IT seeks partly to offset the shortcomings of today's access networks, pending the deployment of fiber to the home. For example, while fixed-line Internet access speeds today do not exceed a few megabits per second, the connection speed of users' PCs can average 100 megabits per second. Fiber optics will align these two speeds while eliminating bottlenecks in the transmission chain.

As of 2007, according to George Gilder, Google, which designs its own datacenter architecture, operated a few dozen datacenters, made up of between half a million and a million computer servers. They are said to have 4 petabytes of memory (RAM) and about 200 petabytes of disk storage, several times what would be needed for the entire visible web in 2007! They receive several hundred million queries per day, implying a data flow of more than one petabyte per second. In the case of Redmond-based Microsoft, it said in May 2008 that it continues to build up its infrastructure, adding roughly 10,000 powerful computer servers a month to its data centers. "It's a staggering amount of computing power, about the equivalent what popular social networking site Facebook uses, according to Chris Capossela [a senior Vice President at Microsoft Corporation]." The Google datacenters are so powerful that they are generally installed in the vicinity of power plants and rivers to keep them powered and cooled at the lowest possible cost. In the world of data centers, the megawatt of electric power is clearly at least as important as the megabyte of computer data. Sources indicate that as of 2007 the five largest search engine companies have a total of 2 million servers amounting to about 600 megawatts, according to the head of operations of Ask.com, Dayne Sampson.[36]

The chairman of Google, Eric Schmidt, believes that mastering datacenters will become a hyper-strategic component in the telecommunications value chain in the years ahead: "We believe we get tremendous competitive advantage by essentially building our own infrastructures." Schmidt goes even further by placing datacenters at the heart of Google's technological strategy: "A third way to think of Google is as a giant supercomputer... There's never been anything like it, so we don't know how to express it. We build our own data centers, and we do a lot of the work ourselves because the commercial solutions do not have high enough performance... we have not only data centers, but we have fiber that interconnects those data centers, and connects to the ISPs. At Google, speed is critical. And part of the way we get that speed is with that fiber." Schmidt goes on to say that among their datacenters "there are a few very large ones. But in a year or two the very large ones will be the small ones because the growth rate is such that we keep building even larger ones, and that's where a lot of the capital spending in the company is going."[37]

The datacenter technology model is driven by the advertising business model. Datacenters are the only constructs today that have the ability and power to manage the multi-dimensional data that describes users' "contexts," as well as to develop "hyper-relevant" applications and services, which are the sources of growth in advertising revenues.

There is ample reason to believe that by 2010 we will see the build-out throughout the world of thousands of datacenters. Those datacenters will represent for the second life of networks what telephone exchanges were for their first life.

This trend toward reintegration of digital resources in the network is known as "Cloud Computing" and has become a whole new area of research and education. Recognizing the shortage of skills in this new area of expertise, IBM and Google, together with six universities, decided in 2007 to create an experimental network in the U.S. This initiative will enable students to acquire the necessary skills for cloud computing and the technology associated with it, known as "virtualization." This will permit management and optimization of resources in the "cloud" in a completely transparent manner for users, who might be using resources originating in a number of different datacenters but appearing as homogeneous.

What constitutes today the full and complete chain of interaction between a user and a datacenter in the case of Google, Microsoft, Yahoo!, Ebay, Amazon and other major Internet players is in fact most of what is necessary for second life networks to operate. Everything from voice to video communications to content discovery, sharing and storage, transmission is there for two or multiple users wherever they are in the world, and in real-time if so needed. The context of the users and the objects they are or may be interested in (written, oral or visual content) can be combined to provide an extremely rich user experience. This will save time as well as allow the rapid discovery and access to information about unknown and relevant content, people, places and opportunities.

Like the "telephone directories" or directory assistance of first life networks, information on people, institutions, and companies are in fact embedded in the fabric of these datacenters. People have become instantly connectable to any other relevant correspondent from within any Internet service they use just by clicking one button. Be it from an email, web-browsing, social networking, gaming, video, or music site.

The inside structure of these private interconnected datacenters shows considerable networking equipments, fiber optics and software based communication features such as Internet voice communication. New networks are defined today by the extensive datacenter silos run by companies like Google, Yahoo!, and Microsoft. It is clear that first life networks gradually morphed into this very powerful combination of datacenters, fiber networks, and content delivery networks. With a major contribution of a new active component of networking: humans. Internet based applications today make decisions and recommendations for users based on their state at different points in time and space. Being always connected (mobile and fixed Internet as one single connectivity paradigm), the datacenters decide the paths, at all levels of the infrastructure, over which the flows of data exchanged travel. It is not hard to imagine how powerful Internet router will be in the near future by applying Moore's law. They will have the wherewithal to make sophisticated decisions continuously. They will in fact retain what happens thru and around them.

Software and hardware will cooperate and have an increasing intimate dependency at the deepest levels of the fabric of second life networks. In addition the techniques of virtualization will become widely used and accelerate the deployment of networks capable of making decisions. Software has become so sophisticated and dense that it is like hardware. Hardware is now virtualized, almost all physical hardware has a virtualization layer on top of it. The next generation datacenter will dynamically configure itself and will automatically, for example, turn off power to its unrequired computing and storage resources and re-activate them when needed.

Since it stores everything and can infer actions it needs to take based on past experience and patterns, the datacenter can make preemptive decisions to increase its overall capacity by sending signals to the appropriate decision-maker, human or machine, to, for example, augment the "hardware side" of bandwidth, storage, and processing power.

Functions until now operating in different logic streams such as transport, routing, storage, search, discovery, browsing, finding, and transacting are increasingly being combined and treated equally across its digital core material. As a result, second life networks will be even more capable to conduct real-time multitasking than their predecessors.

For example when Cisco announced in March 2008 its new generation router Aggregation Services Router (ASR) 1000 Series it immediately referred to the fact it had to custom-build a specific new semiconductor, the QuantumFlow chip: "one of the world's most sophisticated pieces of silicon micro-circuitry," something expected of Intel Corporation or Sun Microsystems Inc. What is interesting is the recognition by Cisco that the new router "needed to run 'stateful' features with no loss of speed. A stateful connection is when a router or other networking device keeps track of an entire packet stream and can recognize which digital packets of information belong with which other packets, where those are supposed to go, and what, exactly, they are supposed to do... The majority of Internet routers until now has been 'stateless' and has treated each packet in isolation.[38]"

There is another interesting inversion emerging today. The mammoth datacenters of the big Internet players, have over time achieved a maturity and size difficult to duplicate. They have a *de facto* comparative advantage akin to those of network operators during networks' first life. These operators were required to "open-up" their infrastructures in well-defined and regulated ways. Today the same issue can be said to apply to these datacenters. These infrastructures it could be argued, which are becoming crucial to the operation of second life networks and store significant information about users' online lives, could do with being opened to other newly formed companies who rely on this data and functionality to succeed.

Ensuring Quality Growth

From a macroeconomic point of view, as was once the case with telephone and high-speed data connectivity, the deployment of new-generation fiber and mobile networks will generate economic growth, wealth, and employment. At a time when developed economies are facing unprecedented demographic changes because of ageing populations, significant investment in networks could be an important part of the solution. Beyond the equipment costs it represents for the value chain of the telecommunications sector, this investment will open up an era of considerable labor productivity for economic sectors that are heavily dependent on information and communications technology. Network bandwidth will experience a leap comparable in scope to that of the late 1990s and early 2000s. It is estimated that information and communications technologies accounted for between 0.3% and 0.9% annual growth in gross domestic product (GDP) in OECD countries during this period.[39] The entry into the core of the second life of networks could result in a similar impact on the growth rates of industrialized countries, somewhere in the range of 0.6% per year (average of the previous range) over the next few years. Increased network bandwidth would have a ripple effect on allied technologies like computer hardware and software, applications, sensors, new forms of software-based media, and the like.

Beyond this estimated average, the impact of growth in each country will, of course, depend on each country's investment effort, the rate at which these technologies are applied, and their adoption by users.

More specifically, this contribution to growth can be broken down into two elements. On the one hand, the investment in equipment required for the deployment of these new networks will have a direct impact on GDP. On the other hand, the use of these new technologies by all economic sectors will foster labor productivity and employment (through the creation of new activities and reintegration into labor markets of senior citizens, people with disabilities, or unskilled individuals), and this will have an indirect but significant impact on growth.

Above all, the economic growth generated by the second life of networks will be primarily high-quality growth, respecting principles of sustainable economic development.

First, as has been the case throughout the 130-year history of networks, the growing substitution of digital connections for physical movement will once again have a significant impact on carbon emissions and thus a positive impact on our environment. The Stern Review report on climate change provides instructive figures in this regard: in the United Kingdom, the telecommunications sector contributes 2.3% of GDP, while devoting only 0.82% of its costs to energy. To use a helpful comparison, telecommunications consumes almost 40 times less energy than air transport per production unit.[40] Thus, according to the 2005 European Commission report "Assessing Opportunities for ICT to Contribute to Sustainable Development," if 50% of the employees in the 25 countries of the European Union were to replace a physical meeting by a teleconference every year, the equivalent of 2.13 million tons of CO_2 would be saved. Similarly, if 20% of business trips in Europe were replaced by a "tele-presence" solution (video or web conference), 22.35 million tons of CO_2 would be saved each year. More precisely, it is only the high speeds offered by fiber that can supply improved and immersive video-conferencing (which requires between 15 and

45 mega bits per second), providing an even more effective substitute for physical meetings.[41]

Second, faster digital exchanges will trigger new developments in the field of health and dependent care. For example, using their mobile phone, patients with chronic diseases (such as diabetes) can continuously transmit biometric data concerning blood sugar to their doctor. Families of Alzheimer's patients would be able to geolocate their loved one so as to ensure the best possible quality of life for him or her. Generally speaking, second-life networks will allow the elderly to be more closely connected to relatives living at a distance. We have already seen strong growth in the consumption of online games among isolated elderly citizens. Third, fiber networks will have wide-ranging societal impacts. Individuals will be able to offload a larger number of necessary but onerous tasks even faster while taking advantage of opportunities to discover information, entertain themselves, and communicate with others more naturally. More active civic engagement will be encouraged through social and recommendation networks. Distance learning will be promoted at all ages through total immersion in knowledge and expertise networks. Government transparency will be increased due to the global dynamics of the "blogosphere," consisting of online forums and widespread generalized collaboration.[42] This will create new social capital as a direct consequence of increasingly fluid exchanges, a development even more important than the transitions from fixed to mobile, mainframe to PCs, telex to the Internet, and finally from broadcast television to cable, satellite, or IPTV.

Vishesh Kumar in his article "Is Faster Access to the Internet Needed?" in the Wall Street Journal on April 10, 2008, notes that: "Verizon and Comcast executives say future technological developments will ultimately vindicate their strategy." He goes on to quote in the article Comcast Chief Executive Brian Roberts who commented at the January 2008 Consumer Electronics Show in Las Vegas that: "Broadband was instrumental to the success of Google, Amazon, eBay, YouTube and all other graphics-and video-rich Web services we now take for granted... So when we boost Web speeds

10, 20, maybe up to 50 times faster than what you're used to today, it will mean a whole new world of innovation that we can barely imagine."[43]

Getting Investors to Invest...

That most of the required technological solutions are available namely, fiber networks, and technologies and mobile transmission protocols such as HSDPA—does not ensure our entry into the core of the second life of networks. In a context of privatization and deregulation of the telecommunications industry, economic conditions, both financial and regulatory, must also be aligned in order for the parties involved to invest in the new networks.

First, from a microeconomic standpoint, the paradigms of the first phase of Internet development and its related services are no longer applicable. Copper networks were well adapted to carrying an abundance of free (or almost free) Internet services and content; they were also generally amortized (although they required incremental improvement and continual maintenance). Fiber networks are new and need to be built from the ground up. To do so, their operators have to gradually find new sources of income so that they can generate an expected return on investment consistent with the amounts invested, the risks involved, and their internal rate of return—that is, the things that define the foundation of any sound business. During the first life of networks, deployment of new access networks required many mechanisms to be put in place in order to transfer income to the operator building the network.

In several countries the connection of residential subscribers to the network was in part financed or subsidized by the higher long-distance rates paid by business users. They were the largest users of this vital resource and, because they absolutely needed it for their business to operate properly, they could not easily reduce their consumption or find a substitute.

The system in most countries was protected by a monopoly concession granted by the state to national operators. After the breakup of AT&T in the U.S. in 1974, local and long-distance services (including billing) were separated, requiring access charges to be introduced so that the long-distance operator would compensate the local network operator for the "last mile" access to users (local traffic was often not measured and was perceived as free because it was part of a flat-fee monthly subscription).

The investment potential of mobile networks, particularly in Europe, has benefited from reductions in regulated tariffs for call termination, which were higher for mobile than for fixed networks. In practice, this means that the share of the price of a fixed-to-mobile call paid back to the mobile network operator by the fixed-network operator is larger than the share of a mobile-to-fixed call paid back to the fixed-network operator by the mobile network operator. In the U.S., mobile phone subscribers pay for the radio resource used in both received and sent calls, which may explain why mobile telephony penetrated more slowly in the U.S. than in Europe. Although the Internet was not built out as an access network from scratch, it represents an interesting case for consideration. In the United States, low-bandwidth access to the Internet was fully handled by local telephone companies, and the data traffic was exempt from "access charges." For broadband, as we have already seen, DSL in the U.S. and Europe was built on existing subscribers' lines. The lines were made available to alternative operators under "unbundling of the local loop" rules, with prices factoring in that those lines were globally amortized, leading ultimately to flat-rate pricing for Internet access. In its early days, the core of the Internet network was funded, in the U.S., by the Department of Defense for its development and by universities for its deployment.

In the radically new context of second-life networks, identification of the new revenue streams that can be generated or captured by network operators in order to justify investment in the building of very high bandwidth networks is the key question. There will certainly be more than one answer, and most probably a range of solutions.

Clearly, the financing of the new infrastructure cannot be fully and directly supported by users, at least not in the build-out phase.[44] Yet network operators may be sorely tempted to question the assumption of neutral access to the Internet (the "Network Neutrality" principle). This principle calls for a network access operator to treat all services and content providers on the Internet equally (i.e., with a flat fee for access to the network), irrespective of differing bandwidth and quality of service required. Questioning this principle opens up the possibility for an operator to differentiate among online service and content providers by charging a rate depending on the differing resource demands made of the network. Thus, Internet content and service providers can be offered different transmission speeds and different levels of quality of service, with higher prices for faster and more reliable communications (in particular, for rich content requiring large amounts of bandwidth and real-time transmission, such as streaming video).

In this model, content and service providers pay the local operator directly for carriage of their information with a certain level of quality. Those who want to benefit from new, advanced features on these networks can do so, but under a different agreement with the operator. This modus operandi is not completely new: during the first life of networks, special business models, such as 800 or toll-free numbers, emerged. With toll-free numbers, the telephone carrier charges the called party, instead of the calling party, for the cost of the calls. This option allowed businesses to provide some services—such as customer care, technical support, and telemarketing—by telephone, by subsidizing all or part of their customers' calls.

Bob Blau, of BellSouth (now AT&T), explained in 2006[45] that his company offered the general public DSL connections at different speeds, including 256 kilobits per second and 128 kilobits per second. But these speeds are insufficient to transmit video. The company Movielink, which offers video on demand (VoD) services on the Internet, had entered into an agreement with BellSouth so that each DSL subscriber who wanted to subscribe to the Movielink VoD service would benefit from an increased speed without having to request it from BellSouth or having to pay an extra charge. In this case, Movielink paid Bell South for the increase in bandwidth.

In the United Kingdom, in 2007, Internet service providers[46] reconsidered their relationships with Internet content and service providers that generated (or were about to generate) a significant level of video traffic on their access network. For example, according to Dan Sabbagh "they want the (BBC) corporation to share the cost of upgrading the network—estimated at £831 million—to cope with the increased workload. Viewers are now watching more than one million BBC programmes online each week [*via* the BBC's iPlayer, Internet video reader software)]. The BBC said ... that its iPlayer service, an archive of programmes shown over the previous seven days, was accounting for between 3 and 5 per cent of all internet traffic in Britain, with the first episode of The Apprentice watched more than 100,000 times *via* a computer."[47]

In general, applications that require a constant and high-quality level of service are likely to buy a special class of broadband connectivity and service from network operators. This is the case for HD video, and also for all real-time applications, such as live TV or telephone. For these purposes, information packets cannot arrive late or be lost, since this might distort the voice of the caller or pixelize the television screen (which can be critical at the moment of a touchdown or a goal). Other applications, from the world of finance or healthcare, for example, are still not fully available on the Internet because they require a service guarantee that current Internet "best effort" delivery methods cannot satisfy. These applications also are likely to require a better quality of service, calling for guaranteed network transmission quality and security, which would justify higher pricing.

CNET reported that: "U.S. telecommunications giant AT&T has claimed that, without investment, the Internet's current network architecture will reach the limits of its capacity by 2010."[48]

According to Jim Cicconi, vice president of legislative affairs for AT&T, "at least $55 billion worth of investment was needed in new infrastructure in the next three years in the U.S. alone, with the figure rising to $130 billion to improve the network worldwide."[49]

However, we must look elsewhere for a significant part of the solution to ensure that operators can get a fair return on their investment in the second life of networks. For example operators themselves can go out and "capture" the value where it is, in order to cover their costs and investments and generate profits. The main contestants so far in the race for online advertising revenues have been the traditional Internet services (such as Google and Yahoo!), but network service operators also have the ability to participate. More generally, those who deploy networks should also share in the value generated by the services they make possible, in order to justify their investments to stockholders. It is not only a question of networks, access, and core, important as they are. It is also about the production of relevant services delivered over these networks. Economically, it is no longer possible to rely solely on the business model of "selling pipes." Instead, network operators must act as true "audience amplifiers." This is a necessary condition so that they can amortize their networks tomorrow, allowing them to invest today.

While sofar network operators have moved only a little in this direction, the content producers—who, coincidentally, are actually in a similar situation—have shown themselves to be more dynamic and innovative. Events such as News Corporation's acquisition of the social networking site MySpace are indicative of this trend, as is the demonstrated propensity of artists themselves to seek new sources of revenues. Examples of such artists are Madonna and Radiohead, both of whom, faced with piracy of their music and the loss of related revenues, decided to use the Internet to sell products and tickets for their public concerts—a way to promote and take advantage of their brand on the Internet. After leaving her record studio, Madonna signed a contract with a concert promoter, Live Nation.

In April 2008, according to the Wall Street Journal, "with her 11th studio album to promote, Madonna is mounting an intimate free concert for fans—and anyone else who wants to watch live online... a day after 'Hard Candy' is released, the pop star will perform at the Roseland Ballroom, a New York club with room for about 3,000 people... (the) concert will be supplemented by a prerecorded interview with Madonna, rehearsals and scenes from backstage. The concert footage will be edited on the fly and streamed *via* Microsoft's MSN network. The show also will be archived[50]".

Radiohead, for their part, terminated their contract with EMI/ Capitol and in October 2007 released on their own their seventh album, *In Rainbows*. Radiohead made the album available on the Internet with no brand and with an "open" sale price: the group simply asked Internet users to pay what they were willing to for the album, saying, "It's up to you," and immediately after, still insisting that "it's really up to you."[51]

In the UK, Prince's new album Planet Earth was provided for free inside the Daily Mail on a Sunday in July 2007[52]. "Rather than release the CD through the traditional channels, he has chosen to give away the album to fans, first, exclusively, free with this newspaper, then to those attending his forthcoming shows in London." The article goes on to quote Prince saying he is "spreading my music and my word to as many people as possible. It's direct marketing, which proves I don't have to be in the speculation business of the record industry, which is going through tumultuous times right now."

Among telecom operators, an interesting and quite exceptional case is that of the Japanese mobile operator NTT DoCoMo (subsidiary of the incumbent carrier NTT). The company has successfully undertaken a strategy of dynamic vertical integration. NTT DoCoMo is a pioneer in its core business, particularly in the area of bandwidth: in mobile 3G (with nearly 40 million subscribers in early 2008, according to the company), in 3G+, and soon perhaps in 4G, with packet transmission speeds of as much as 5 gigabits per second[53] during tests! Next, NTT DoCoMo climbed to the upper layers of the value chain in services and in content. For example, the company established a partnership with Google in early 2008 to develop multiple uses of mobile Internet, defying Yahoo! which was positioned ahead of Google in Japan. It also positioned itself strongly in the "m-payment" domain, as well as those of RFID and M2M. NTT DoCoMo even entered the realm of content by acquiring a 5% stake in Fuji TV. The company's strategy relies on significant R&D spending, ranging between 2% and 2.5% of its turnover.[54]

Of course, all these economic considerations of players' strategies to capture value fall within a regulatory framework that is decisive with regard to the required investment decisions. From this point of view, three key issues are emerging. First, the colossal investment required for the construction of fiber-optic access networks requires a stable playing field for the players of the value chain. Second, contrary to the entire history of access network deployments (including Internet access and "unbundling" of copper pairs), both existing and alternative operators are now on an equal footing with regard to investments in fiber optics[55]—a situation that calls for more symmetrical regulation. Third, and more generally, the changes in the telecommunications value chain with the entry of new players—services and content—more than ever requires dynamic and adaptive regulation, in line with the new situation.

In this regard it is interesting to note the recent statement (April 2008) from the chairman of the Federal Communications Commission, Kevin Martin, who said he opposed other network-opening proposals that would undermine investment incentives. "This careful balancing of spurring innovation and consumer choice while encouraging infrastructure investment is critical to the wireless industry's continued impressive growth."[56]

When Europe and Asia will Wake up

In macroeconomic terms, there is another vital element for worldwide entry into the second life of networks. As long as a European or Asian Google does not emerge, the transfer of advertising revenues from Europe and Asia to the U.S. should accelerate at the same pace as the growth of revenues generated by advertising on the Internet and, increasingly, on mobile. For Europe, this transfer could grow from approximately $7 billion in 2007 to about $20 billion in 2012.[57] In Asia, over the same time period, it would grow from $6 billion to about $15 billion.[58]

This is not a criticism. This reflects a simple fact: the creative genius of a company like Google is its ability to develop services that bring value and benefits to Internet users throughout the world. The challenge for Europe and Asia is to reach at least the same level of creativity and efficiency as Silicon Valley, to train and mobilize for this purpose armies of engineers and entrepreneurs specializing in advanced software development, particularly in the context of Cloud Computing. Young Asians and Europeans are no less innovative than young Californians. Indeed, many successes in Silicon Valley are the result of ideas of people coming from Europe, China, India, and other countries. Above all, they should not try to catch up, by inventing a pale substitute for existing services provided by Silicon Valley companies. Quite the contrary, they must come up with something visionary—the "next big thing." The sum of all these innovations from different parts of the world can only accelerate the advent of new, truly disruptive technologies relevant to improving the quality of the worlds' economy and as a result the well being of its people.

The reason for the success of Google and other American information technology service companies also lies in their massive technological capital. Those who believe that Google, for example, develops only on the basis of the ideas of a handful of engineers are clearly mistaken. Google is a formidable and gigantic information factory. Alongside the development of their human capital, Europe and Asia must rapidly participate in the building of powerful, next-generation networked datacenters, the "server farms" needed to accelerate the computing power necessary to provide hyper-computing power, responsiveness, and relevance worldwide.

The case of Africa is special. Fixed networks there are still undeveloped, and as a result mobile networks are often the only type of network available. Users are leapfrogging first-life networks and going directly to their second-life manifestations. Some African countries, where the penetration of fixed telephone and banking infrastructures is thin, could be among the first to benefit from "m-payment" for services—that is, payment by mobile. Of course, most African countries do not play on the same playing field as the industrialized countries, but this leap forward represents a real opportunity for their economic development.

Protecting Personal Data

Entering the second life of networks depends on satisfying another major condition: the protection of users' personal data. The development of services and advertising on the highly personalized second-life networks will quickly generate a desire by users for greater privacy and security in the way their personal data is used. The great discontent of the members of the social network Facebook in November 2007, after they realized that their personal profile and even their online purchases had been used for advertising purposes and shared with their social networks, demonstrates the significance of privacy protection. The immediate negative outcry over the intrusion into users' lives was a wake-up call for Facebook, as we have seen. In March 2008 its management apologized to its users and withdrew the feature called Beacon, which was responsible for the situation. The company subsequently put in place explicit new rules[59] that increased privacy options. Following this up in May 2008, Facebook increased online safety on its service by putting in place 40 safeguards to protect young people from sexual predators and cyber-bullies. MySpace, had already agreed in January 2008 to implement safeguards on its social network service.

The concept of personal data includes several elements, which form the "profile" of the user. The first category of personal information relates to individual identity: name, surname, nickname(s), date of birth, mailing address, telephone number(s), address(es), e-mail(s), key and password(s), bank account information, and so on.

With the growth of second-life networks, user identities have multiplied at an increasing rate.

The second category of personal information is demographic. This is information that describes an individual's or household's income level, age, ethnicity, regional location, number of children, goods owned, and services used. This is the type of information traditionally gathered in polls or national censuses.[60]

The third category of personal information includes data called "context" or "attention." This type of data has a temporal dimension (for example, the web sites visited in the past, tracked by the well-known "cookies" placed on users' computers) and a spatial dimension (the physical location determined by GPS or other means).

The confidentiality of personal data is an important concern for all members of the telecommunications value chain, limiting the extent to which they can unilaterally disclose or use the data of their clients (except in the case of judicial decree, of course). It aims to protect the privacy and even intimacy of users who, for example, may not want their friends to know where they are or what they buy. Confidentiality in networks must be user-defined above all, as a function of the degree of "openness" accepted by the user. However, levels of confidentiality can vary greatly. For example, elder Internet users may be very reluctant to give out information about themselves. In contrast, the friendly and relaxed atmosphere that often prevails on social networks for young people seems to encourage participants to disclose their personal data more promiscuously.

A study conducted in August 2007 by the U.S. company Sophos, a specialist in computer security, of a representative sample of Facebook users highlights these new patterns: 41% of users disclose intimate personal (and sometimes confidential) information, like their email, birth date or phone number, to completely unknown users. For example, as part of an experiment conducted for this study in which users responded to a fictional character, 23% of respondents disclosed their real phone number, 26% their instant messaging nickname, 78% their mailing address, 84% their birth date, and 87% details of their vocational training or company for which they work.[61]

In general, the level of privacy afforded by the service provider is often lower than the actual level of confidentiality demanded by the user. This discrepancy stems from a lack of transparency and simplicity in the terms and conditions of service that are provided to users, and in particular the clauses regarding the use of their personal data. Users often accept, for example, the terms and conditions of the services to which they subscribe without reading them (are they even designed with the objective of being read?). In some social networks, acceptance of terms and conditions gives implicit permission for the service to mine the user's address book and send emails soliciting his or her contacts, without necessarily informing clearly the user.

These practices are often criticized for the lack of transparency of their "opt-out" clauses for the user when they attempt to use default personal data. Usually, the user discovers the situation by accident and vigorously objects. This could be avoided by using an explicit "opt-in" clause that requires the prior consent of the user to exploit personal data. Beginning in 2008, this issue generated particular interest and dialogue following the acquisition of DoubleClick by Google.[62]

Basically, it is as if users are at the junction of two separate groups of infinitely meshed networks: on one hand, a group of "aware" networks, with which they communicate deliberately and with full knowledge and control of their own actions; and on the other hand, a group of "unconscious" networks extracting information about them and then using it without their knowledge, or at least without much transparency. The day is coming when users will require the establishment of a shelter, a form of protective insulation for these unconscious networks to which they are connected. Alternatively, the migration of unconscious networks to aware networks through a transparent policy of opt-in and opt-out might fix the problem. The provider that can securely provide this shelter will certainly gain an edge over other players. For operators and service providers, the ability to ensure the protection of personal data will surely be a key differentiating attribute.

Protecting privacy
"Heavens! It's my husband!"
"One day I am going to find Google in the closet!"
Source: Chimulus, *La Tribune*, November 8, 2007.

Ensuring the confidentiality of personal data is not a simple matter in the second life of networks, which is characterized by a total immersion of users in networks, in which the users' trails are fairly easily to trace. However, confidentiality is not incompatible with the wealth of new services, even those that use contextual data, such as geographical location. Once companies are committed to greater transparency, there can be a trade-off between protection and use of personal data, ensuring a satisfactory degree of privacy in exchange for the benefits provided by these services.

One can imagine, for example, that users might choose profiles or "personas" that summarize a set of specific behaviors, which would then guide and determine the services and content offered to them on networks. This concept of hyper-configurability would allow for a certain degree of anonymity in conjunction with a simple way of indicating choices and tastes (compared to an online form requiring item-by-item selection of what data elements they are willing to allow a service to use).

Beyond privacy, security of personal data must of course be ensured. This concept has more to do with the protection of sensitive data that might be diverted and used for fraudulent purposes. Security is provided by increasingly sophisticated information-encrypting systems on the networks.

Today many people know how to allow Internet "cookies" from certain sites they trust to be installed and kept on their PC. It is likely that a good compromise will gradually be found between offering a conscious opt-in or opt-out decision procedure to users and maintaining an acceptable level of comfort in the day-to-day use of Internet services.

What Future for the Value Chain?

The projection of the telecommunications value chain globally over the short to medium term suggests two main conclusions. First, the network operator layer is likely to remain the most significant in terms of value added, even if its growth rate is the lowest in the value chain. Second, the layer expected to be the most dynamic is the services layer, followed by the content layer.

Regarding the positioning of players, in addition to fiber deployment and the attendant entry into the core of networks' second life, the value chain will continue its fusioning and attain an equilibrium where each player's comparative advantage will determine the extent of its territory. The resulting situation could thus resemble "silos" where any given player could control several elements of the value chain.

For their part, at least some network operators will join the movement initiated by players in the other three layers. With new revenues being gradually preempted by newcomers, they will need to stake out a place at the table for themselves. Having evolved to a role of audience amplifiers, they will know that in the race to develop and provide service and content they have key comparative advantages, such as their proximity to and knowledge of the client. Furthermore, their ability to guarantee the security of personal data while responsibly maximizing the constellation of value propositions they can offer users is a key competency of networks in their second life.

*

Beyond the stakes represented by the physical investments that will be a key factor for the success of the second life of networks, human capital will make the difference in the end. People will play a major role at the heart of this transformation, not only as nodes of these networks but also through the new abilities their varied professional experiences, skills, and cultural diversity have allowed them to forge. The melding of younger generations all across the value chain, will incorporate their capacity to generate and use innovations, drive unimagined inventions, and energize the telecommunications sector. Those who will win will be those who listen, observe, understand and are willing to interrupt their linear and incremental thinking in order to incorporate these constant streams of new ideas and behaviors.

These people whose talents were forged in the crucible of networks' first life will have an advantage as they stand on the brink of a new frontier: that of the second life of networks.

Once again: we are the networks.

Bibliography

ANDERSON, Chris, 2006, *The Long Tail: Why the Future of Business is Selling Less of More* (New York: Hyperion).

BARABASI, Albert-Laszlo, 2003, *Linked: How Everything is Connected to Everything Else and What It Means* (New York: Plume).

BATTELLE, John, 2005, *The Search: How Google and Its Rivals Rewrote the Rules of Business and Transformed Our Culture* (New York: Penguin-Portfolio).

BECK, John C., and Mitchell WADE, 2004, *Got Game: How the Gamer Generation Is Reshaping Business Forever* (Cambridge, Mass.: Harvard Business School Press).

BENKLER, Yochai, 2006, *The Wealth of Networks: How Social Production Transforms Markets and Freedom* (New Haven: Yale University Press).

BERNERS-LEE, Tim, Mark FISCHETTI, and Michael L. DERTOUZOS, 2000, *Weaving the Web: The Original Design and Ultimate Destiny of the World Wide Web by Its Inventor*. (New York: HarperCollins).

BRAFMAN, Ori, and Rod A. BECKSTROM, 2006, *The Starfish and the Spider: The Unstoppable Power of Leaderless Organizations* (New York: Penguin-Portfolio).

CHRISTENSEN, Clayton M., Erik A. ROTH, and Scott D. ANTHONY, 2004, *Seeing What's Next: Using Theories of Innovation to Predict Industry Change* (Cambridge, Mass.: Harvard Business School Press).

FLACHER, David, Hugues JENNEQUIN, and Jean-Hervé LORENZI, 2007, *Réguler le secteur des télécommunications?—Enjeux et perspectives* (Paris: Economica).

FRANSMAN, Martin, 2007, *The New ICT Ecosystem: Implications for Europe* (Edinburgh: Kokoro).

FRIEDMAN Thomas L, 2005, *The World Is Flat: A Brief History of the 21st Century* (New York: Farrar, Straus, and Giroux).

GLADWELL, Malcolm, 2005, *Blink: The Power of Thinking without Thinking* (New York: Little, Brown).

GLADWELL, Malcolm, 2002, *The Tipping Point: How Little Things Can Make a Big Difference* (New York: Little, Brown).

HAMEL, Gary, and Bill BREEN, 2007, *The Future of Management* (Cambridge, Mass.: Harvard Business School Press).

HEATH, Chip, and Dan HEATH, 2007, *Made to Stick: Why Some Ideas Survive and Others Die* (New York: Random House).

MARKOFF, John, 2005, What the dormouse said: How the sixties counterculture shaped the personal computer industry (New York: Penguin Group).

METCALFE, Robert M, 2000, *Internet Collapses and Other InfoWorld Punditry* (Boston: IDG Books Worldwide).

RAYMOND, Eric S, 2001, *The Cathedral and the Bazaar: Musings on Linux and Open Source by an Accidental Revolutionary* (Sebastopol, California: O'Reilly Media).

RHEINGOLD, Howard, 2003, *Smart Mobs: The Next Social Revolution* (New York: Perseus).

STEPHENSON, Neal, 1992, *Snow Crash* (New York: Bantam Books).

SUROWIECKI, James, 2004, *The Wisdom of Crowds* (New York: Anchor Books).

TAPSCOTT, Don, and D. Anthony WILLIAMS, 2006, *Wikinomics: How Mass Collaboration Changes Everything* (New York: Portfolio).

VON HIPPEL, Eric, 2005, *Democratizing Innovation* (Cambridge, Mass.: MIT Press).

Notes

Introduction

1. Netscape Communications (formerly Mosaic Communications Corporation) is an American company created in 1994, which developed and marketed the Internet browser called Netscape Navigator.

2. Backbone networks are the essential highways of information that connect the different networks of the Internet. If we make the natural analogy to roads, they are the highways that connect two separate cities (networks). Distribution and access networks play the role of a city's road and driveways that allow a person to drive their car home or to receive truck deliveries.

3. By analogy with the use of this term in Raymond, Eric S., 2001, *The Cathedral & the Bazaar: Musings on Linux and Open Source by an Accidental Revolutionary* (Sebastopol, California: O'Reilly).

4. http://www.newsweek.com/id/130285

CHAPTER 1
Once Upon a Time, in Networks' Earlier Life...

1. http://www.stern.nyu.edu/Sternbusiness/spring_summer_2002.pdf (page 44).

2. http://inventors.about.com/od/bstartinventors/a/telephone.htm

3. http://inventors.about.com/od/bstartinventors/a/Alexander_Bell_2.htm

4. http://en.wikipedia.org/wiki/Photophone

5. Sawyer, W. E., 1880, "Seeing by Electricity," *Scientific American*, 42, p.373.

6. Langrand, Patricia, 2005, "Réseaux de télécommunication et services de contenus : l'addition de forces complémentaires au service du client," *La Jaune et la rouge*, 604, (Paris : École Polytechnique), pp. 57-62.

7. The radio was invented in the 1890's by Guglielmo Marconi. His idea was to send telegrams to ships, in the form of Morse code, without using wire (i.e., by means of radio signals). He patented his invention in the UK in 1897. This is ano-

ther example of research designed to improve telecommunications (by telegraph) leading to a new disruptive technology. Indeed, in France for example, radio was originally called "wireless telegraph."

8. http://memory.loc.gov/ammem/edhtml/edcyldr.html

9. http://www.webbconsult.com/1800.html

10. Dr. Moses Greeley Parker had been so convinced of the potential of the telephone that he reportedly acquired a large number of shares in the American Telephone Company and the New England Telephone and Telegraph Company and in 1883 would become their largest individual shareholder. (http://susning.nu/hist.html and http://www.grantsmanagement.com/parkerhistory.html).

11. According to Park D., *Good Connections: A Century of Service by the Men & Women of Southwestern Bell,* Southwestern Bell Telephone Company, Strowger first got suspicious when upon the death of a close friend, the departed's family contracted with his competitor for the funeral services.

12. Fischer, Claude S. 1994, *America Calling: A Social History of the Telephone to 1940*, (Berkeley and Los Angeles: University of California Press).

13. http://www.privateline.com/TelephoneHistory5/partyline.htm

14. http://www.recherche-innovation.equipement.gouv.fr/IMG/pdf/Les_telecommunications_cle7272fc.pdf

15. Within the ITU, the ITU-T is in charge of standardization matters.

16. It is often said that the birth and subsequent spectacular development of the Silicon Valley in California owe much to the Defense Research carried out not only during the World War II, but also during the Cold War that immediately followed and subsequently as part of the race to militarize space (cf. Ronald Reagan's "Star Wars" presentation, March 23, 1983).

17. http://inventors.about.com/od/rstartinventions/a/radio_2.htm.

18. The "modem" modulates a frequency according to the digital signal generated by the computer or the subscriber (i.e., converts it into an analog transmission signal) or "demodulates" the analog signal coming from the traditional subscriber line (i.e., converts it into a digital signal compatible with digital networks).

19. In Greek mythology, the Minotaur was a creature that was part man and part bull. It dwelt at the center of the Labyrinth, which was an elaborate maze-like construction built for King Minos of Crete and designed by the architect Daedalus and his son Icarus who were ordered to build it to hold the Minotaur. (http://en.wikipedia.org/wiki/Minotaur).

20. http://en.wikipedia.org/wiki/Defense_Advanced_Research_Projects_Agency

21. The packets of IP networks, called "datagrams," are different in nature from those of X.25 packet networks, in which a single identifier connects packets of a same communication and therefore all use the same path, called a "virtual circuit."

22. "Broadband Connectivity, Competition Policy," FTC Staff Report, June 2007, p. 14.

23. Ring D., 1947, "Mobile Telephony: Wide Area Coverage," Bell Laboratories Technical Memorandum. (http://www.sri.com/policy/csted/reports/sandt/techin2/chp4.html).

24. http://en.wikipedia.org/wiki/Martin_Cooper

25. Micro electronic chip technology encrypts the identity of the subscriber and shields it at its most vulnerable, that is, during the communication between the telephone and the network. This also protects direct reading of the information from the chip as any such attempt simply destroys the chip.

26. http://www.gsmworld.com

27. "Thriving in Harmony: Frequency Harmonisation: the Better Choice for Europe," UMTS Forum, November 2006, p. 14.

28. Proust, M., 1983, *Remembrance of Things Past*, Volumes I-III. (New York: Vintage Books). For an interesting essay on the relationship of Proust's characters to technology see: http://culturemachine.tees.ac.uk/Cmach/Backissues/j001/articles/art_naas.htm

CHAPTER 2
The Gestation of Second Life Networks

1. http://www.washingtonpost.com/wpdyn/content/article/2007/05/23/AR2007052301418_pf.html

2. This was not the first collaboration between Jobs and Wozniak. In 1971, they created the "bluebox," which generated network tones that could bypass the US telephone system's central office-based billing systems of the time, thereby allowing free calling. (http://www.markusehrenfried.de/mac/applehistory.html).

3. http://en.wikipedia.org/wiki/Breakout

4. April 2008. (http://news.netcraft.com/archives/web_server_survey.html).

5. http://www.comscore.com/press/release.asp?press=2018

6. The "HTTP," or Hypertext Transfer Protocol used on the Internet is itself a complete communication language. On the Internet, as indicated earlier, each packet or datagram is independent. It is very different from voice communications where each call requires the previous call to be explicitly suspended. HTTP allows users the possibility with each click to connect to a document on a different server than the one currently engaged (that is, without having to hang up and redial a new).

7. It would be eventually adopted by Microsoft for its "Explorer" software.

8. The Opera browser was developed and launched in 1994 as a research project of Telenor, the Norwegian telecommunications operator; it has since been adapted to mobile devices as well as gaming consoles.

9. http://www.thecounter.com/stats/2007/November/browser.php

10. http://www.internetworldstats.com/stats.htm

11. http://www.internetnews.com/feedback.php/http://www.internetnews.com/dev-news/article.php/3718896

12. http://en.wikipedia.org/wiki/Wintel

13. The Personal Digital Assistant is a device the size of a small blackboard slate which includes a planning calendar as well as basic office applications. In 2007, a number of smartphones including the Apple iPhone have the same size, if not smaller, as well as the same functionalities as a traditional PDA.

14. http://www.internetworldstats.com/blog.htm

15. http://news.netcraft.com/archives/2008/01/28/january_2008_web_server_survey .html

16. Coffman K.G., and A.M. Odlyzko, 2002, « Growth of the Internet », *Optical Fiber Telecommunications IVB: Systems and Impairments* (I. Kaminow, T. Li (éds), Academic Press), pp. 17-56 and Rosston G., 2006, *The Evolution of High-Speed Internet Access: 1995-2001*, Stanford Institute for Economic Policy Research Discussion Paper, pp. 05-19. (http://papers.ssrn.com/sol3/papers.cfm?abstract_id=943570# PaperDownload).

17. http://www.pewinternet.org/pdfs/Internet_Status_2005.pdf

18. http://www.websiteoptimization.com/speed/tweak/average-web-page/

19. Thomas L. Friedman, 2005, *The World Is Flat: A Brief History of the Twenty-First Century* (New York: Farrar, Strauss and Giroux).

20. Litan R., and A. Rivlin, 2001, *The Economic Payoff from the Internet Revolution*, Brookings Institution, Washington, DC, cited *in* Atkinson R., and A. McKay, 2007, *Digital Prosperity*, The Information Technology & Innovation Foundation (ITIF).

21. http://www.sciam.com/article.cfm?id=000B0C22-0805-12D8-BDFD83414B7F 0000&page=2

22. NPD Group Inc., cited *in* Darlin D., 2007, "The Hard Drive as Eye Candy," *The New York Times*. (http://www.npd.com/press/releases/press_070102.html)

23. http://en.wikipedia.org/wiki/Wirth's_law

24. http://www.softwaremetrics.com/Articles/history.htm

25. According to Wikipedia: "In the economic literature, a value chain, or chain of activities, can be defined as a set of products (goods and services) and producers that cooperate to serve a market." (http://fr.wikipedia.org/wiki/Cha%C3% AEne_de_valeur).

26. Equipment manufacturers had already begun to sell private equipment directly to enterprises, notably the Private Automatic Branch Exchange systems or PABX.

27. Actually, the user did not have to pay for the Minitel terminal, which remained the property of the telecoms operator.

28. The service did not support pictures, sound, or video, however.

29. http://www.sandman.com/telhist.html et http://som.csudh.edu/fac/lpress/471 /hout/telecomHistory

30. http://www.neweconomics.org/gen/uploads/gx2dcv45szxpsn55yohg3rjh0108 2003160019.pdf

31. "The Future of Video: New Approaches to Communication Regulation", in Report of the 21st Annual Conference on Telecommunication Policy, Aspen Institute, August 16-19, 2006.

32. O. Brafman, R. Beckstrom, *The Starfish and the Spider: The Unstoppable Power of Leaderless Organizations*, (New-York, Penguin-Portfolio, 2006).

CHAPTER 3
We Are the Networks

1. Term first used by Tim O'Reilly, author, publisher and founder of O'Reilly Media, Incorporated.

2. It is interesting to note the reaction attributed to Bill Gates by David F. Marquardt, a member of his board of directors, who puzzled by the fact that Microsoft had done little in the Internet space: "His point of view was that the Internet was free: how could that be an interesting business"? A short time after that, Microsoft bought part of the Internet network provider UUNET and then, in 1997, invested one billion dollars in the cable company Comcast, convinced that controlling the telecommunications network was an essential part of an Internet strategy.

3. Kathy Rebello, with Amy Cortese and Rob Hof, "Inside Microsoft (Part 2)", Cover Story Business Week, via http://www.businessweek.com/1996/29/b34842.htm, (accessed June 8, 2008).

4. "An Estimated 43 Billion Text Messages Were Sent Globally on New Years Eve", cellular-news, posted January 9, 2008, http://www.cellular-news.com/story/28496.php. Dusan, "Gartner: Next year mobile phones users in major matkets will send more than 2 trillion text messages", posted December 19, 2007, via http://www.gart ner.com/DisplayDocument?ref=g_search&id=5507070&subref=simplesearch, (accessed June 8, 2008). http://www.intomobile.com/2007/12/19/gartner-next-year-mobile-phones-users-in-major-markets-will-send-more-than-2-trillion-text-messages.html, (accessed June 8, 2008). Full report: Nick Ingelbrecht et al, "Market Trends: Mobile Messaging, Worldwide, 2006-2011," Gartner, November 26, 2007, http://www.gartner.com/DisplayDocument?ref=g_search&id=550707&subref=simplesearch

5. In 1970, AT&T began selling through its equipment subsidiary Western Electric> the first videophone under the name "Picture-Phone." This device, conceived in 1956, was publicly shown for the first time at the New York World's Fair in 1964.

6. Amanda Lenhart, Mary Madden, Alexandra Rankin Macgill and Aaron Smith, "Teens ans Social Media", PEW/internet: PEW Internet & American Life Project, http://www.pewinternet.org/pdfs/PIP_Teens_Social_Media_Final.pdf

7. Ted Cohen, President of Tag Strategic, *in* "A Report of the 15th Annual Aspen Institute Roundtable on Information Technology 2007." http://www.aspeninstitute.org/atf/cf/%7BDEB6F227-659B-4EC8-8F848DF23CA704F5%7D/NETGENERATION.PDF

8. Linda Stone, "Thoughts on Attention and Specifically, Continous Partial Attention", Linda Stone, http://www.lindastone.net.

9. All these services were provided by the country's incumbent carriers of the time except for Prodigy, a subsidiary of IBM and Sears & Roebuck.

10. When the Internet first appeared, a service was created that allowed text based electronic messaging communications to be sent to the Internet from a Minitel.

11. "Stanley Milgram", Wikipedia, http://fr.wikipedia.org/wiki/Stanley_Milgram.

12. One example in the US is imbee.com (for parents, teachers, and children).

13. Facebook-Press Room, "Applications", http://www.facebook.com/press/info.php?statistics.

14. For references to the mathematical theory of graphs, see http://fr.wikipedia.org/wiki/Theorie_des_graphes.

15. As is the case with Facebook or the majority of other social networks, there exist a number of "gateways" in Second Life that allow to link the user's virtual and real social networks, through, for example, SMS's that can be sent bi-directionally.

16. A "Maven" is someone with considerable influence on other members of a social network. The role of Mavens in the propagation of knowledge has been recognized in a number of domains, from politics to social trends. (http://en.wikipedia.org/wiki/Maven).

17. Malcom Gladwell, *The Tipping Point: How Little Things Can Make a Big Difference*, (New York, Little, Brown and Company, 2002).

18. In thermodynamics and statistical physics, entropy is a magnitude that quantifies the state of excitation, or "disorder," of a system, which is synonymous with heat or agitation. The second principle of thermodynamics postulates that all physical transformation of a system creates entropy. Sugar melts in coffee progressively in a disordered fashion. There is almost no chance of it resuming its crystalline form spontaneously.

19. Stone, http://www.lindastone.net

20. David P. Reed, "The Law of the Pack" *Harvard Business Review*, February 2001, p. 23-24: "[E]ven Metcalfe's Law understates the value created by a group-forming network [GFN] as it grows. Let's say you have a GFN with n members. If you add up all the potential two-person groups, three-person groups, and so on that those members could form, the number of possible groups equals 2n. So the value of a GFN increases exponentially, in proportion to 2n. I call that Reed's Law. And its implications are profound."

21. See http://en.wikipedia.org/wiki/Dunbar's_number for references and a summary.

22. Jean M. Twenge, *Generation Me: Why Today's Young Americans Are More Confident, Assertive, Entitled—and More Miserable Than Ever Before*, (New York: Simon & Schuster, 2007).

23. J.D. Lasica, "The Mobile Generation: Global Transformations at the Cellular Level », report of the 15th Annual Aspen Institute Roundtable on Information Technology 2007".

24. Deloitte & Touche USA LLP, "2007, State of the Media Democracy," 2007, via http://www.deloitte.com/dtt/article/0,1002,cld%253D182990,00.htm

25. John C. Beck, Mitchell Wade, *Got Game: How the Gamer Generation Is Reshaping Business Forever*, (Cambridge (United States): Harvard Business School Press, 2004).

26. Tomi Ahonen and Alan Moore, "Cyworld is like MySpace 2 years into future says Herald Tribune", September 27, 2006, Communities Dominate Blogs, http://communities-dominate.blogs.com/brands/2006/09/cyworld_is_like.html

27. A high school student in the 15-17 age group is quoted as saying: "Email has become obsolete. MySpace is much faster. It is like sending an SMS on your phone. You can also send pictures."

28. http://www.pewinternet.org/pdfs/PIP_Teens_Social_Media_Final.pdf

29. Sean McGrath, "it's not what you know, but who you know", ITworld, February 16, 2007, http://www.itworld.com/Tech/2987/nlsebiz070220/index.html and James Governor, Redmonk, http://redmonk.com/jgovernor/

30. James Surowiecki, *The Wisdom of Crowds*, (New-York: Anchor Books, 2004).

31. It is important that these other sites be as independent as possible in order to avoid the real problem of the "lemming effect" or "Panurge's sheep." Panurge, a character in Francois Rabelais' Pantagruel, extracts revenge on a merchant by buying one of his sheep and leading it off a cliff, all of the merchant's other sheep followed suite. The concept is used to describe the blind following of perceived authority. It does no good to consult multiple sites if they are influenced by each other and propagate, in particular, an inaccurate piece of information. The fact that they may sometime be accurate is not a basis for reliability. See *Gargantua and Pantagruel* by François Rabelais, Thomas Urquhart, Peter Anthony Motteux, Barnes & Noble, January 2005.

32. Voltaire, article "Art dramatique," *Questions sur l'Encyclopédie*, 1764 (citation in Italian).

33. "WikiScanner", Wikipedia, http://en.wikipedia.org/wiki/WikiScanner, and Emily Biuso, "Wikiscanning", New York Times, December 9, 2007, via nytimes.com. http://www.nytimes.com/2007/12/09/magazine09wikiscanning.html?_r=1&oref=slo gin &pagewanted=print

CHAPTER 4
Free as a Business Paradigm

1. "Jean-Baptiste Colbert", Wikipedia, http://en.wikipedia.org/wiki/Jean-Baptiste_Colbert.

2. This is not necessarily always the case anymore.

3. "Potlatch", Wikipedia, See "Potlatch", Wikipedia, http://en.wikipedia.org/wiki/Potlatch.

4. The term "communalism" is often used instead of "communism" as a way of referring to communal societies that are not based on Marxism. See "Communalism", Wikipedia, http://en.wikipedia. org/wiki/Communalism, and Kenneth J. Gergen, An invitation to Social Construction, (London, Thousands Oaks, and New Dehli: SAGE Publications, 1999), http://books.google.com/books?id=ygIiu6Vx7kC&printsec=titlepage &dq=technocommunalism&source=gbstocs&cad=1

5. "Google and IBM Announce University Initiative to Address Internet-Scale Computing Challenges", Google Press Center, October 8, 2007, http://www.google .com/intl/en/press/pressrel/20071008_ibm_univ.html

6. National Science Foundation, Global Environment for Network Innovations http://www.geni.net/

7. Kathy Rebello, with Amy Cortese and Rob Hof, "Inside Microsoft (Part 2)", Cover Story, Business Week, via Business Week http://www.businessweek.com/1996/29/b34842.htm

8. *Ibid.*

9. Bought in April 2005 by Adobe.

10. Premium-rate telephone number, Wikipedia, http://en.wikipedia.org/wiki/Premium-rate_telephone_number.

11. Michael Arrington, "Yahoo Mail Annonces Unlimited Storage", Tech Crunch, March 27, 2007, http://www.techcrunch.com/2007/03/27/yahoo-mail-announces-unlimited-storage/.

12. Chris Anderson, "Free! Why $0,00 Is Future of Business", Wired, February 25, 2008, http://www.wired.com/techbiz/it/magazine/16-03/ff_free?currentPage=all, (accessed June 8, 2008).

13. *Ibid.*

14. Jorge Cauz "Collaboration and the Voices of Experts", Britannica, June 3, 2008, http://britannicanet.com/

15. Source: TiVo, "Trend Sheet for GAAP Statement of Operations", http://files.shareholder.com/downloads/TIVO/263371852x0x177668/a1212103-09eb-47d7-a591-eb95ea76578e/38825A.PDF

16. Maria Aspan, "TiVo Shifts to Help Companies It Once Threatened", New York Times, Published: December 10, 2007, Via Nytimes.com
http://www.nytimes.com/2007/12/10/technology/10tivo.html

17. The users' locations are given by either satellites or by the cells defining the wireless network (the latter is achieved by triangulation of the transmitters that register the presence of the mobile telephone in a given network cell). The so-called Assisted GPS accelerates the identification of location by using the location of the base station to which the mobile phone is connected to augment the satellite's calculation.

18. Source: PwC/IAB Internet Advertising Revenue Report (http://www.iab.net/about_the_iab/recent_press_releases/press_release_archive/press_release/64544), and Zenith Optimedia, Advertising Expenditure Forecasts, December 2007.

19. User figure is for Q1, 2008

20. eMarketer, This chart is developed from the eMarketer report titled US Online Advertising: Resilient in a Rough Economy, March 2008. Link only available by subscription. See http://www.emarketer.com/SiteSearch.aspx?arg=US+Online+Advertising+Spending%2c+by+Format&src=search_go_sitesearch

21. The decision in "MGM v. Grokster" handed down by the US Supreme Court on June 27, 2005 held illegal the distribution and promotion of peer-to-peer file sharing software because it infringes the intellectual property and copyright laws by sharing music without compensating its rights holders).

22. Jefferson Graham, "Napster's back to basic: Free tunes", USA Today, May 1, 2006, via http://www.usatoday.com/money/industries/technology/2006-04-30-napster-free_x.htm

23. "Music Licensing", Wikipedia, http://en.wikipedia.org/wiki/Music_rights see also "Digital Millennium Copyright Act" Wikipedia http://en.wikipedia.org/wiki/digital Millenium Copyright Act

24. "Nokia World 2007: Nokia outlines is vision of Internet evolution and commitment to environmental sustainability", Nokia, December 4, 2007, http://www.nokia.com/A4136001?newsid=1172937; also see: David Chartier, "Nokia's

Comes With Music' now comes with EMI", Ars Technica, March 20, 2008, http://arstechnica.com/news.ars/post/20080320-emi-to-come-together-with-nokias-comes-with-music.html

25. Vauhini Vara, "Companies Tolerate Ads to Get Free Software," *The Wall Street Journal*, March 27, 2007, via http://online.wsj.com/public/article/SB117496231973149939-05bQvBXmI0tpwoBlv0d98dW4r70_20080325.html).

26. "Spiceworks", SandHill.com, June 2, 2007, http://www.sandhill.com/community/company_spotlight.php?id=23

27. Victoria Colliver, "Medical site is on a mission to set records", San Fransisco Chronicle, March 16, 2007, via SFGate, http://www.sfgate.com/cgibin/article.cgi?f=/chronicle/archive/2007/03/16/BUG 9OOM1FJ1.DTL&hw=practical+fusion&sn=1000.

28. Software 2007 conference Proceedings, May 8-9, 2007, http://software2007.com/grafix/pdf/Enterprise-Software-Customer-Survey-2007.pdf

29. Benjamin J. Romano, "EA boss calls in-game ad revenues forecasts 'widly high'", Seattle Times, November 29, 2007, via http://blog.seattletimes.nwsource.com/techtracks/archives/2007/11/ea_boss_calls_ingame_ad_revenue _forecasts_wildly_h.html

30. Saul Hansell, "How Big is Google? Here's Another Measure", The New York Times, April 18, 2008, via http://bits.blogs.nytimes.com/2008/04/18/how-big-is-google-heres-another-measure/

31. Independent Television (generally known as ITV) is a public service network of British commercial television broadcasters, set up under the Independent Television Authority (ITA) to provide competition to the BBC. ITV is the oldest commercial television network in the UK. See "ITV", Wikipedia see http://en.wikipedia.org/wiki/ITV, (accessed June 8, 2008).

32. Millward Brown, "Brandz top 100 brands shows dramatic growth in the financial power of brands", April 8, 2008, http://www.millwardbrown.com/Sites/Optimor/Media/Pdfs/en/Brandz/Brandz-2008-PressRelease.pdf

33. "David Ricardo", Wikipedia, See http://en.wikipedia.org/wiki/David_Ricardo

CHAPTER 5
A Premonitory Turmoil

1. Mortgage loans made to high-risk, non-"prime" borrowers. The growing number of foreclosures during the summer of 2007 triggered a worldwide destabilization of the financial sector.

2. "Structural" changes are long-term movements having fundamental impacts as opposed to "cyclic" variations that are short-term and tend to have imperceptible impacts on the fundamentals of a sector. The fact that the telecommunications sector does not seem to suffer from the cyclical slowing in the global economy suggests that deeper structural forces sustain it.

3. This was not AT&T's first antitrust case. In 1949, the US government tried to force AT&T to divest itself of Western Electric, its equipment arm. The final decision in 1956 while not going as far as divestiture did require AT&T to limit its

activities to traditional telecommunications services as well as to open, through licensing, its patents (including those of Western Electric) to third parties.

4. http://www.cfo.com/article.cfm/3005316/c_2984343/?f=archives

5. That is, an over-cable TV delivery network as opposed to an over-the-air TV network.

6. This is done by publicly offering to buy the company's stock for a price per share usually reflecting a premium over the stock exchange price.

7. WorldCom obtained worldwide notoriety with its accounting scandal of 2002.

8. Lucent was the post-divestiture subsidiary formed by AT&T in 1996 that combined the old Bell System subsidiaries Western Electric (equipment) and Bell Labs (Research & Development).

9. Hewlett-Packard, "Merger & Acquisitions Announcements", http://h30261.www3.hp.com/phoenix.zhtml?c=71087&p=irol-mergers

10. Justin Scheck and Ben Worthen, "Hewlett-Packard Takes Aim at IBM", *The Wall Stree Journal*, via http://online.wsj.com/article/SB121067909289288201.html?mod=us_business_whats_news

11. Scientific Atlanta is a company that has been around longer than Cisco, but whose skill set has become crucial given the new developments in audiovisual media (videos, HD TV, etc.) over fixed networks.

12. The PC, historically considered part of information technology value-chain, became "communicating" with the advent of the Internet, and in so doing has become associated with the equipment layer of the telecommunications value chain.

13. May 2008.

14. These include Target, the NBA, Sears Canada, Sears UK, Benefit Cosmetics, Bebe Stores, Timex Corporation, Marks & Spencer, Mothercare and Lacoste to name a few. (http://en.wikipedia.org/wiki/Amazon.com).

15. Among these web services are included Amazon Simple Storage Service (Amazon S3), Amazon Elastic Compute Cloud (Amazon EC2), Amazon Simple Queue Service (Amazon SQS), Amazon Mechanical Turk (MTurk) and Amazon E-Commerce Service (ECS), renamed Amazon Associates Web Service. (http://www.amazon.com/gp/browse.html?node=3435361).

16. Larry Dignan, "Amazon's cloud computing will surpass its retailing business", ZDNet, April 14, 2008, http://blogs.zdnet.com/BTL/?p=8471

17. Quentin Hardy, "The Death of Hardware," *Forbes*, February 11, 2008 (http://www.forbes.com/forbes/2008/0211/036.html).

18. "Ebay Developer's Program", Ebay, http://developer.ebay.com/

19. "NCR Corporation", Wikipedia, http://en.wikipedia.org/wiki/NCR_ Corpo ration

20. "Microsoft Invests $200 Million in Qwest; Qwest to Deliver Complete Line of Electronic Commerce, Enterprise Network and Managed Software Solutions based on Windows NT Server", Microsoft Press Release, December 14, 1998, http://www.microsoft.com/presspass/press/1998/Dec98/QwestPr.mspx

21. In France, as of the end of the 1990's, the Vivendi group, already owning at least in part Cegetel, SFR, and Canal+, attempted to "verticalize" its part of the telecommunications value-chain by purchasing in 2000 the Seagram Company, owner of Universal Studios, for about $35 billion.

22. Robert Andrews, "Nokia Says 'Less A Manufacturer, More An Internet Company'; Plans Share Buy-Back", *The Washington Post*, May 8, 2008, via http://www.washingtonpost.com/wp-dyn/content/article/2008/05/08/AR2008 050801655.html

23. Dawn C. Chmielewski and Alex Pham, "Sony to launch video service for PlayStation 3", *The Los Angeles Times*, April 21, 2008, http://www.latimes.com/business/la-fi-sony21apr21,1,152898.story

24. "Garmin® nüvifone™ Takes Personal Navigation and Communication to the Next Level", Garmin, January 30, 2008, http://www8.garmin.com/pressroom/mobile/013008.html and http://www8. garmin.com/nuvifone/

25. "SFR Signs Partnership Agreement with Archos, Leader in Portable Multimedia Players", (translated from french), Joint SFR/Archos press release, February 12, 2008, http://www.archos.com/corporate/press/press_releases/ARCHOS_SFR_FINAL _20080212.pdf

26. It may come as a surprise to some readers that we consider Microsoft to be part of the telecommunications value-chain. Microsoft is present in the equipment layer (with software like Internet Explorer designed to allow a simple and graphic access of hardware to the network as well as its "communicating" Xbox gaming console), but also in the service and content layers due to its MSN Internet portal and online software. The traditional core competencies of Microsoft (symbolized by Windows) would put it only in the neighboring computer value chain.

27. Generation Partner, http://www.generation.com/

28. "Google Will Apply to Participate in FCC Spectrum Aution", Google press release, November 30, 2007, http://www.google.com/intl/en/press/pressrel/fccspectrum_20071130.html

29. Matt Ritchel, "Technology Group Plans Wireless Network", *The New York Times*, May 7, 2008, http://www.nytimes.com/2008/05/07/technology/07sprintweb.html?_r=1&oref=slogin

30. Dionne Searcey and Kevin J. Delanay, "Google, Goldman ANd Hearst Invest In Broadband Firm", *The Wall Street Journal*, http://online.wsj.com/article/SB112069176731378945.html

31. François Sterin, "About the Unity bandwidth consortium", The Official Google Blog, February 25, 2008, http://googleblog.blogspot.com/2008/02/about-unity-bandwidth-consortium.html

32. More precisely, Akamai is not in the business of building a true network infrastructure but rather an "overlay" network that will ensure the quality of real-time content streams such as voice and video over the Internet.

33. "Nokia to extend leadership in enterprise mobility with acquisition of Intellisync", Nokia press release, November 16, 2005, http://press.nokia.com/PR/200511/1021663_5.html

34. The name given sometimes to the "Net Generation," made up of individuals born between the mid-1970's and mid-1990's, who are keen users of a wide range of new technologies.

35. Mark Borden et al., "The World's Most Innovative Companies," *Fast Company*, March 2008. (http://www.fastcompany.com/magazine/123/the-worlds-most-innovative-companies.html).

36. Thomas Friedman, *The World Is Flat: A Brief History of the Twenty-First Century*, (New York: Farrar, Strauss, and Giroux, 2005).

37. Eric S. Raymond, *The Cathedral & the Bazaar: Musings on Linux and Open Source by an Accidental Revolutionary*, (Sebastopol, California: O'Reilly, 1999).

38. Adam Lashinsky, "RAZR'S edge", *Fortune*, June 1, 2006, via http://money.cnn.com/2006/05/31/magazines/fortune/razr_greatteams_fortune/index.htm

39. Ken Polsson, "Chronology of IBM Personal Computer", Ken Polsson, April 22, 2008, http://www.islandnet.com/~kpolsson/ibmpc/

40. Richard Kalgaard, "The Cheap Revolution" Forbes. See also http://www.forbes.com/forbes/2003/0428/037_print.html

41. Of course, Mark Zuckerberg is reminiscent of another Harvard student who actually interrupted his studies to found a company: Bill Gates.

42. Jim Collins, jimcollins.com, http://www.jimcollins.com/

CHAPTER 6
Outlook:
Entering the Core of Networks' Second Life

1. http://www.forrester.com/Research/Document/Excerpt/0,7211,42496,00.html

2. http://www.itu.int/ITUD/icteye/Reporting/ShowReportFrame.aspx?ReportName=/W/MainTelephoneLinesPublic&RP_intYear=2006&RP_intLanguageID=1

3. New York Spa offers "Black Berry Thumb" Massage, http://www.foxnews.com/story/0,2933,251963/00.html

4. MMayja Palmer and Paul Taylor, "Google Homes in on Revenues from Phones," *Financial Times*, 13 February 2008. (http://www.ft.com/cms/s/667f13de-da60-11dc-9bb90000779fd2ac.html).

5. http://www.wired.com/gadgets/wireless/magazine/16-02/ff_iphone

6. "The Paperless Map is the Killer App", http://www.businessweek.com/magazine/content/07_48/b4060067.htm

7. http://www.businessweek.com/magazine/content/07_48/b4060067.htm

8. The term comes from George Gilder. (http://www.businessweek.com/the_thread/dealflow/archives/2005/05/into_the_gilder.html).

9. According to Verisign, which manages the worldwide addressing system of RFID radio-tags, in *The Economist*, 26 April 2007.

10. http://www.itu.int/osg/spu/publications/internetofthings/

11. That is, 100 billion, billion, billion, billion.

12. http://www.pewinternet.org/PPF/r/232/report_display.asp

13. http://www.pewinternet.org/pdfs/PIP_Online_Video_2007.pdf

14. This includes only downloads and not "streamed" videos—that is, content that is watched but not saved on the user's hard disk. Streamed video is growing faster; see "Reuters Information Week," December 27, 2007.

15. http//www.nowpublic.com/tech-biz/youtube-10-hours-content-every-minute

16. According to Quincy Smith, president of CBS Interactive, at the "Newteevee" conference in San Francisco, fall 2007.

17. http://www.twice.com/article/CA6465724.html

18. http://www.wired.com/techbiz/it/news/2008/04/broadband2

19. http://www.itu.int/osg/spu/publications/internetofthings/

20. The laws of Moore, Cooper, Reed...

21. 100,000 e-mails (with no attached file) of average size amount to 1 gigabyte; 1 gigabyte can also accommodate 350–500 digital mid-resolution pictures and roughly an equivalent amount of MP3 or WMA music.

22. Source: http://newsroom.cisco.com/dlls/2008/ekits/IP_Traffic_Chart_030408.pdf

23. "The Coming Exaflood," *Wall Street Journal*, 20 January 2007 (http://online .wsj.com/article/SB116925820512582318.html).

24. http://www.llrx.com/features/deepweb2007.htm

25. http://www.emc.com/leadership/digital-universe/expanding-digital-universe.htm

26. Nate Anderson. Ars Technica. http://arstechnica.com/articles/culture/the-coming-exaflood.ars/2

27. http://www.usatoday.com/tech/products/services/2008-04-20-internet-broad band-traffic-jam_N.htm

28. Bruce Mehlman was assistant secretary of commerce under President Bush. Larry Irving was assistant secretary of commerce under President Bill Clinton.

29. Bruce Mehlman. http://www.washingtonpost.com/wp-dyn/content/article/2007/05/23/AR2007052301418_pf.html

30. http://en.wikipedia.org/wiki/Parkinson's_law

31. http://en.wikipedia.org/wiki/Jean-Baptiste_Say

32. It is always necessary to be cautious with spectacular predictions, for they can be hazardous. One of the more famous instances is that of Bob Metcalfe. In an article in the 4 December 1995 issue of the magazine *Infoworld*, he predicted that in 1996 there would be a catastrophic collapse of the Internet, which he called "gigalapse" (a contraction of "giga" and "collapse"). He recognized his mistake by eating his words, literally (by cutting his article into little pieces and eating these pieces of paper) during his closing speech at the 6th conference on the World Wide Web in Santa Clara, California, in 1997. (http://www.merit.edu/mail.archives/nanog/1997-04/msg00192.html).

33. http://blogs.sun.com/jonathan/entry/moving_a_petabyte_of_Data.

34. http://en.wikiquote.org/wiki/Andrew_S._Tanenbaum (cited in *Computer Networks*, 4th ed., p. 91).

35. http://www.akamai.com/html/about/press/releases/2007/press_082707.html

36. George Gilder, "The Information Factories," *Wired*, 2006, 14.10. (http://www. wired.com/wired/archive/14.10/cloudware.html).

37. http://www.wired.com/techbiz/people/news/2007/04/mag_schmidt_trans

38. http://startupschool.org

39. "Measuring the impacts of Information and Communications Technologies (ICT) across official statistics," OCDE, Working group on indicators for the information economy, January 2008.

40. "The Stern Review Report on the Economics of Climate Change," Nicolas Stern for the UK government, 2006, chapter 11: "Structural Change and Competitiveness," pp. 15–16. (http://www.hm-treasury.gov.uk/media/0/9/Chapter_11_Structu ral_Change_and_Competitiveness.pdf).

41. "The new technologies of information and communication in supporting sustainable development?", *La note de veille*, The Center of Strategic Analysis, October 2007, vol. 78, p. 3.

42. California Broadband Task Force, January 2008 (http://www.calink.ca.gov/taskforcereport/).

43. http://online.wsj.com/public/article_print/SB120779422456503907.html

44. However, in early 2008, Time Warner Cable wondered about its strategy of unlimited Internet access and tested a pay as you go scheme for consumption beyond a certain threshold. The problem they faced in terms of capacity was mainly in the consumption of streamed VoD. "Time Warner Cable has indicated that 5% of its customers consumed 50% of its network capacity."

(http://query.nytimes.com/gst/fullpage.html?res=9B0DEED7123CF932A15752C0A96E9C8B63&sec=&spon=&pagewanted=2).

45. "The Future of Video: New Approaches to Communication Regulation," The Aspen Institute Communications and Society Program, twenty-first annual Aspen Institute Conference, 16–19 August 2006.

46. "ISPs fear that introduction of web broadcasts will overload their networks as users download 'catch-up' TV."

Source: The Independent (http://www.independent.co.uk/news/business/news/internet-groups-warn-bbc-over-iplayer-plans-461167.html).

47. http://technology.timesonline.co.uk/tol/news/tech_and_web/article3716781.ece

48. http://www.news.com/ATT-Internet-to-hit-full-capacity-by-2010/2100-1034_3-6237715.html

49. *Ibid.*

50. http://online.wsj.com/article/SB120915827211445737.html

Music: Seeing Madonna Free By John Jurgensen The Wall Street Journal.

51. http://www.time.com/time/arts/article/0,8599,1666973,00.html

52. http://www.dailymail.co.uk/pages/live/articles/showbiz/showbiznews.html?in_article_id=468443&in_page_id=1773 and http://www.dailymail.co.uk/pages/live/articles/live/live.html?in_article_id=466634&in_page_id=1889

53. http://www.phonearea.net/2007-02-10/ntt-docomo-achieves-worlds-first-5gbps-packet-transmission-in-4g-field-experiment/

54. Citigroup, "Successful Vertically Integrated Telecom Models Are Exceptional: DoCoMo in Japan," 19 September 2006.

55. Except for civil engineering issues.

56. http://www.reuters.com/article/technologyNews/idUSN0126806320080401?feedType=RSS&feedName=technologyNews

57. Calculation based on Zenith Optimedia, Advertising Expenditure Forecasts, December 2007.

58. *Ibid.*

59. http://online.wsj.com/article/SB120586883223146093.html

60. See http://www.census.gov/dmd/www/content.htm for information regarding the U.S. Census.

61. http://www.sophos.com/pressoffice/news/articles/2007/08/facebook.html

62. http://newsinfo.inquirer.net/breakingnews/infotech/view/20080123-114225/Internet-privacy-concerns-cause-very-public-r

Table of Contents

CHAPTER 2

The Gestation of Second Life Networks

CHAPTER 3

We Are the Network

CHAPTER 4

Free as a Business Paradigm

CHAPTER 5

A Premonitory Turmoil

CHAPTER 6
Outlook:
Entering the Core of Networks' Second Life